Mobile Learning through Digital Media Literacy

new literacies ¶
AND DIGITAL EPISTEMOLOGIES

Colin Lankshear and Michele Knobel
General Editors

Vol. 73

The New Literacies and Digital Epistemologies series
is part of the Peter Lang Education list.
Every volume is peer reviewed and meets
the highest quality standards for content and production.

PETER LANG
New York • Bern • Frankfurt • Berlin
Brussels • Vienna • Oxford • Warsaw

Belinha S. de Abreu

with Vitor Tomé

Mobile Learning through Digital Media Literacy

PETER LANG
New York • Bern • Frankfurt • Berlin
Brussels • Vienna • Oxford • Warsaw

Library of Congress Cataloging-in-Publication Data

Names: de Abreu, Belinha S., author. | Tomé, Vitor, author.
Title: Mobile learning through digital media literacy / Belinha S. de Abreu, Vitor Tomé.
Description: New York: Peter Lang, 2017.
Series: New literacies and digital epistemologies; vol. 73 | ISSN 1523-9543
Includes bibliographical references and index.
Identifiers: LCCN 2016030612 | ISBN 978-1-4331-2895-0 (hardcover: alk. paper)
ISBN 978-1-4331-2894-3 (paperback: alk. paper) | ISBN 978-1-4539-1903-3 (ebook pdf)
ISBN 978-1-4331-3780-8 (epub) | ISBN 978-1-4331-3781-5 (mobi)
Subjects: LCSH: Mobile communication systems in education. | Internet in education.
Computers and literacy. | Digital media.
Classification: LCC LB1044.87 .D42 2017 | DDC 371.33—dc23
DOI 10.3726/978-1-4539-1903-3
LC record available at https://lccn.loc.gov/2016030612

Bibliographic information published by **Die Deutsche Nationalbibliothek**.
Die Deutsche Nationalbibliothek lists this publication in the "Deutsche Nationalbibliografie"; detailed bibliographic data are available on the Internet at http://dnb.d-nb.de/.

29 Broadway, 18th floor, New York, NY 10006
www.peterlang.com

DEDICATION

To my Dad, José Jesus de Abreu, the person who most inspired my global interests, who enjoyed so much learning from my travel adventures and telling people about my globetrotting. Your presence in our lives is missed.

—SUA FILHA, BELINHA

CONTENTS

FOREWORD

Kat Stewart

Senior Director, NCTA Foundation

As a science fiction fan, I remember the first time I saw Han Solo being frozen in carbonite. It was terrifying and fascinating. I wondered if Han would be okay, but, more importantly, what would happen to the Rebellion while he was in stasis. In the year he was frozen, his world changed. After waking, it took time for him to catch up and understand all that had happened while he was asleep. Him playing catch-up reminds me how quickly things change here on Earth. Imagine if I took a 20-year nap in carbonite. I would awake today to a completely different world. The internet has evolved to be our main source of information and communication. Instead of flip phones, there are smartphones. Not that people use the phone to talk to each other, they text instead. Flying cars aren't a thing, but self-driving ones are real. I'll have to learn all about social media and start taking pictures of my lunch. And what happened to the Spice Girls?

With all these changes, I'll need to figure out how to participate in the digital world. Getting my hands on a mobile device is relatively easy, as is downloading apps and jumping online. Good thing those are designed to be used intuitively. (Once I learn how to swipe, of course.) The big challenge will be understanding how to act and behave while using these digital media platforms. It's the same challenge that students are faced with today. While the world has changed dramatically in the last 20 years, schools have mostly remained the same. That's why this book's look at media literacy education and mobile learning is so important.

For more than a decade, my work has focused on the intersection of media, technology, and education with special emphasis on digital citizenship education, including medial literacy skills, and 21st-century readiness for every student. I've had a ring-side seat to watch the evolution of technology and the internet. As I was developing a standards-based curriculum to teach digital citizenship concepts, Belinha De Abreu was at the top of the list to appear in the curriculum's videos as a content matter expert to explain the tectonic shift that's taking place. With her tenure as a teacher at the K–12 level, as well as the university level, Belinha is uniquely able to temper academic theory with the reality of classroom learning and lead this book's exploration of mobile learning through a media literacy lens.

Belinha and I have presented together at numerous conferences, including SXSWedu in Austin, Texas, on many topics such as media literacy, privacy, and digital citizenship. We have long discussions about the role of technology in education, and she's taught me to look at things from a teacher's perspective. It should be all about the students; all about their learning. Students' lives revolve around mobile devices, with many getting a smartphone or tablet in elementary school, so it's time to meet the students where they are and leverage the devices they use everyday.

The experience of mobile is different from sitting at a computer or in front of a TV. The way you interact with mobile devices is a more intimate experience than when using a mouse— touching, swiping, using your fingers to zoom in to dive deeper into the content—it's much more active and interactive. The app developers are crafty in their use of risk and reward, enticing the user to attempt new things, to not be afraid to fail, and to keep trying until you solve the puzzle, unlock the next level, or gain a prize. You're given an achievable goal; it's hard to reach, but you have many incentives to keep trying until you make it. And this doesn't just happen in games; this type of risk and reward is also evident in social media, encouraging you to post a picture or status that will garner the most likes or more followers. It's also seen in apps or on websites that call for activism or joining friends to support a cause. Because the content is designed to provoke specific behavioral responses, using mobile devices in education increases the possibility of capturing the students' attention and motivating them to act. This potential is discussed throughout the book, particularly in Chapter 3.

Since mobile content is different, teaching and learning with these devices must also be different. It's not enough to know how to make the app work; by bringing in media literacy concepts, students can begin to understand that someone made the world that lives within the app. When a fifth-grader is swimming with whales inside a virtual reality simulation, she is probably in awe of her watery surroundings, the giant creatures she's observing, and the unfamiliar sounds she's hearing. But as this student applies her media literacy skills, she'll keep in mind that virtual reality is not actually reality and may only offer a limited or skewed view of the whales and their habitat. This simulation is a great starting point to learning about whales. Apps and website such as this can be wonderful teaching tools. They let the teacher be creative when tackling a topic and can ignite curiosity in students, motivate them to dig deeper, and encourage them to create content of their own. With these platforms, students also have the potential to make a huge impact on the world. The ice bucket challenge, for example, started as a small campaign and grew to be a worldwide phenomenon that raised millions of dollars for research. Videos going viral, ideas being shared, or funds being raised, the possibilities with mobile are only limited by your imagination. As this book applies a media literacy framework to using mobile for learning, it also talks about the value of creativity when using the mobile content.

While mobile technologies and learning have many benefits, there are certain considerations that give me, and many others, pause. Data privacy, for instance, is such a hot topic an entire chapter of this book is dedicated to it. Now I'm perfectly fine with trading some of my information for a more personalized experience, but I do this as an educated consumer who takes the time to review policies and understand what data is being used and how. Students need to understand this too. They're using mobile at an impressive scale and doing remarkable things with it. They're creating content, and not just videos that they hope will go viral; they're also writing code to make their own apps. They're constantly connected with friends and with people from around the world. They're also active participants in communities that support their passions and work together to effect change. It's an exciting time. And it makes the discussions in this book about how to teach students to be mindful about their data and how to manage their digital footprints in smart and effective ways even more important.

So much is happening in the mobile space—devices are becoming smaller and more affordable, schools are increasing their bandwidth and access, and the number of apps available is increasing exponentially. Mobile has made life easier and will continue to change the world in unexpected ways. I'm glad to see education is beginning to leverage the devices and applications students are using every day. To move this needle further, it's important to consider how media literacy can offer a framework for the development of mobile learning in the classroom. This book plays a significant role as catalyst for the next step in this journey and makes me excited to see what happens in the next 20 years.

PREFACE

Matteo Stocchetti

Principal Lecturer, PhD,

Department of Culture and Communication,
Arcada University of Applied Sciences

When Belinha asked me to write the preface for her new book the first question that came to my mind was: does anybody still read prefaces nowadays?

The digital age, among other things, is the age of performativity (knowledge appreciated for its relation with power instead of truth) and speed—the speed of light that governs the composition, decomposition, and re-composition of reality in the many aspects of digitization.

The reality that gets hammered out by the combination of speed and performativity has violent connotations that can be appreciated if one thinks that the difference between a slap and a caress is precisely speed, and that performativity is a fundamental dimension in the distinction between independent and corporate research.

The elements of violence associated with the enforcement of the digital turn in education and elsewhere are overlooked only because we are getting used to them. Powerless knowledge and the knowledges of the powerless are suppressed by the need of performative knowledge that is all the more urgent the more the technocratic management of society fails to deliver its promises. Reflective processes that require slowness in intellectual work, in interpersonal relations, in teaching, etc., are increasingly challenged and sometimes vanishing because slowness itself has become rare and subversive: a luxury and a risk accessible to a select few. In the digital age, speed is a feature of the technological infrastructure enforced on "human resource." In this development, the "goal"—people—becomes a tool—resources—in the preservation of a socio-political order, anachronistically, but still fundamentally, based on the myths of technological progress. The adaptation of human processes to the speed of the machinery is only one aspect in this process of substitution: a sign that instrumental reason ultimately rules the logic of late capitalism. Slowness violates this subordination, violates the logic of instrumental reason, and is therefore experienced as an exclusive privilege or an act of subversion.

In these circumstances, reading prefaces—as you are doing now—becomes a small act of subversion because it challenges both performativity and speed. It challenges performativity because you will not find a summary of the book in this text. It also challenges speed because it slows you down and, perhaps, it takes your attention somewhere else. But it is also an act in some measure defining social identities: the relative position of an individual as defined by the coordinates of symbolic and cultural capital in society.

The second thought I had after Belinha's kind request was that perhaps there was a point that she wanted me to make for those among us indulging in the luxury of reading prefaces—and de facto, if not in principle, rejecting the performativity and the incitation to speed of the digital age. I spent a fair amount of time thinking about what this point could be. What is it that should be written for those who can take the time and have the curiosity to read what most would consider safe enough to skip?

The social identity or relative positioning of the reader connects with Belinha's book in at least two aspects. The first and rather explicit one is the inherent ambivalence of mobile technology in education. The second aspect, more implicit, is about the shift from the logic of knowledge to the logic of power in the resolution of this ambivalence in educational practices.

Ambivalence and Indeterminacy

One of the main points of the book is that the impact of mobile technology on education is ambivalent: it all depends on intervening variables that vary from learners' "meta" skills to a complex bundle of conditions that can only be generically subsumed under the label of "socio-economic background."

High achieving students, for example, are not affected by the use of mobile phones at school but low achieving ones are. Self-regulating skills, among others, seem to be the intervening variable that explains differences in outcomes.

What this means in practice is that mobile technology in education can do good or bad, depending on who is the user—fair enough. In the politics of education, however, this also means that the disambiguation of technological affordances depends on the social identity of the learners (who s/he is) and on the quality of her education.

Now the problem is that too much of digital/mobile affordances are discussed as if there were no such intervening variables. The "democratic" myth of techno-centric discourse hides the discriminating effects of technological development: more often than not the powerful are usually in a better position than the powerless to take advantage of these affordances.

In uncritical discourses this ambivalence is construed not as an aspect that should invite caution but as a rhetorical device or "trick" to divert responsibility and blame from corporate propaganda to the individual, her community, and/or to educational institutions.

On theoretical grounds, this ambivalence suggests attention to intervening variables that, while perhaps constituting the social identity of the users, are nevertheless, for the most part, beyond his/her control. On these grounds the problem is to understand when and how, for example, mobile "smart" phones can enrich the educational experience or simply superimpose the "consumer" on to the "learner."

On performative grounds, however, responsibility for outcomes, functional or dysfunctional, stays with the users, very much like in the case of the automobile. In the case of underage users, responsibility stays with the relevant adults, usually parents and teachers. The relative position of these agents in the disambiguation of technological indeterminacy is a pretty uncomfortable one. If and when mobile technology fails in education, and the individuals cannot be blamed, then they are the culprits.

The argument of corporate propaganda—the educational impact of mobile technology is what the users make of it—looked at from the point of view of social identities, thus becomes a performative statement that contributes to the neoliberal project in at least two ways. Firstly by construing a community of isolated individuals and secondly by deploying the conceptual infrastructure for the selective distribution of blame according to a familiar pattern: private virtues vs. public vices, profits are private but costs are public, etc.

Enforcing ambivalent technology, denying the theoretical, moral, and, ultimately, political relevance of technological indeterminacy, helps the reproducing and enhancing of social inequalities. It is perhaps time to acknowledge that the problem of inequality of impact is more relevant to the problem of unequal access—the "digital divide." In its mainstream formulations, the latter reflects the myth of technological egalitarianism (the false belief that technology alone can eliminate social differences and inequalities), promotes the "progressive" image of corporations in the digital business, and legitimates their growing profits and influence in the organizations of our societies.

Forcing educational institutions to embrace technology, to embrace the cause of the "digital divide" while disregarding the inequality of impact, means to force the same institutions into accepting the discriminatory effects associated with ambivalence *and* the ideology supporting them: the neoliberal project of forming isolated individuals and the strategy of managing technological risk in our "risk societies" by blaming families and education.

To shift the focus of the discussion from the digital divide to the inequality of impact means to discuss the educational role of mobile technologies in relation to its empirical effects and risks rather than its mythical representations.

And this is where performativity and speed enter the game to affect its results. The rhetoric of "revolutionary change" and a pressing "digital future," carrying inescapable and merciless consequences for those who resist it, transforms the problem of understanding what is happening into the problem of enforcing an uncritical adaptation to a state of affairs.

As I shall argue in a moment, by transforming a problem of knowledge into a problem of power, this shift invites the politicization of education as a legitimate intellectual response.

From the Logic of Knowledge to the Logic of Power

Another point that emerges quite clearly from Belinha's book is the individual and collective efforts made by committed teachers to integrate mobile technologies into meaningful educational practices—in our terms, to resolve the inherent ambivalence of these technologies and to reduce its risks. From the point of view of the politics of education, the strategy that more or less explicitly inspires these efforts is usually called "creative adaptation" or "productive resistance." This strategy can be important and influential in the specifics—given classrooms in given socio-economic contexts—but can hardly challenge the cultural influence of the technocratic myths (Mosco, 2004) as these are adopted by the Neoliberal project and deployed extensively in the "mediatization" of reality.

If one is serious about the educational and pedagogical challenges posed by the digital performativity of this project, the implications of the digital turn cannot be looked at out of context, as if digital technology and learning would be united in a political vacuum. Rather, they should be considered in relation to the politics of education as a fundamental dimension of the competition for the control of the social construction of reality at a local and global level.

More mature or experienced readers may recall, for example, the notion of "context control" proposed by Jean Françoise Lyotard (1979/1982) to describe the distinctive and violent form of legitimation gained through the performative control of reality:

> This procedure operates within the following framework: since "reality" is what provides the evidence used as proof in scientific argumentation, and also provides prescriptions and promises of a juridical, ethical, and political nature with results, one can master all of these games by mastering "reality". That is precisely what technology can do. By reinforcing technology, one "reinforces" reality, and one's chances of being just and right increase accordingly. ...This is how legitimation by power takes shape. Power is not only good performativity, but also effective verification and good verdicts. It legitimates science and the law on the basis of their efficiency, and legitimates this efficiency on the basis of science and law. It is self-legitimating, in the same way a system organized around performance maximization seems to be. Now it is precisely this kind of context control that a generalized computerization of society may bring. The performativity of an utterance, be it denotative or prescriptive, increases proportionally to the amount of information about its referent one has at one's disposal. Thus the growth of power, and its self-legitimation, are now taking the route of data storage and accessibility, and the operativity of information. (p. 47)

In this perspective, Belinha's case for coupling the introduction of mobile learning technologies with media literacy is a discussion not merely about the *methods* but, more radically and rightly in my view, about the *purposes* of education.

As Neil Postman (1993) argued over 20 years ago, "A new technology does not add or subtract something. It changes everything" (p. 18). Technological change in education is always "systemic": a change not only in "how" we do things but also in "what" we do and, especially, "why" we do it. In this change, the "medium is the message" as MacLuhan famously put it or, more prosaically, the tool becomes the end.

The discussion about the purposes of education is a political discussion to the extent that different purposes are sought and legitimized by different participants in relation to implicit or explicit ideas about the future of society. The invitation to discuss purposes rather than merely methods should lead to the realization that the digital turn in education is not a problem of *knowledge* anymore, but rather a problem of *power*.

The ideological implications and risks of the digital turn in education have been known for some time now, perhaps since the late '80s, when books such as *The Myth of the Information Revolution* (Traber, 1986) or *The Ideology of the Information Age* (Slack & Fejes, 1986) were published in response to the early wave of uncritical enthusiasm for the "information revolution." More recently, and despite having been discredited by a number of studies (Helsper & Eynon, 2009; Serazio, 2015; Smith, Skrbis, & Western, 2012), the performative representation of generational differences in terms of a binary pitting "digital natives" against "digital immigrants" (Prensky, 2001; Tapscott, 1998) is still a prominent motive inspiring the enforcement of the digital turns in education. Key notions in the discourse of democratic education such as "freedom," "equality," "empowerment," etc. (Freire, 1974/2013; Giroux, 2004) are hijacked and redeployed in the promotion of digital learning (Mautner, 2005; Newson, 2004) in support of technocentric tendencies in education (Fernedding, 2003). And when governments of countries such as Finland are still willing to invest shrinking educational budgets in mobile technology despite that empirical evidence that "students who use computers very frequently at school do a lot worse in most learning outcomes" and "that technology is of little help in bridging the skills divide between advantaged and disadvantaged students" (OECD, 2015, p. 3), on which grounds can one argue the question is still about (the lack of) knowledge rather than power?

When reality is not anymore the *environment* but the *stake* of political competition, and the logic of scientific enquiry is forced to comply with the criteria of performativity, the idea that one could rely on empirical research to find or formulate optimal solutions is either naïve or manipulative. It is naïve because it ignores the practical impact of performativity. It is manipulative because it hijacks the authority of science to serve performative functions.

When research is inspired by speed and performativity, the role of science is not anymore that of establishing truth as the intersubjective grounds for the mediation of competing interests in society. Rather, it becomes that of enforcing reality itself—the reality of the interests that control the production of knowledge against the "utopia" of alternative interests and reality.

The nature and direction of the ideas and forces associated with the uncritical endorsement of digital technology in education are quite clear (McChesney, 2013). The question is how to oppose these ideas and resist these forces: of rejecting the Neoliberal project and the influence of its prophets in education.

On intellectual grounds the challenge is quite formidable and requires a profound change in the responsibilities associated with the pedagogical work of professional educators. We need a quantum leap in educational discourse, a shift in the way we conceive the nature and purposes of education from a "technical" to a "political" view or, in sociological terms, from a "traditional" to a "critical theory" of the digital turn in education. And, to start with, we don't need to reject technology but only its uncritical usage in education.

I believe that among those who still read prefaces one can find what Postman (1993) called "loving resistance fighters," or perhaps simply critical thinkers with an inclination for action. My invitation for these is: please continue to read this book, learn how the story goes even when the ending is not as happy as we would like it to be. Learn and think how to make sure that the goals are not determined by tools.

References

Ferneding, K. (2003). *Questioning technology. Electronic technologies and educational reform.* New York, NY: Peter Lang.

Freire, P. (2013). *Education for critical consciousness.* London: Bloomsbury (Original work published 1974)

Giroux, H. A. (2004). Critical pedagogy and the postmodern/Modern divide. *Teacher Education Quarterly*, Winter, 31–47.

Helsper, E., & Eynon, R. (2009). Digital natives: Where is the evidence? *British Educational Research Journal*, 1–18.

Lyotard, J.-F. (1982). *The postmodern condition: A report on knowledge.* Minneapolis, MN: University of Minnesota Press (Original work published 1979)

Mautner, G. (2005). The entrepreneurial university. A discoursive profile of a higher education buzzword. *Critical Discourse Studies*, *2*(2), 95–120.

McChesney, R. (2013). *Digital disconnect. How capitalism is turning the Internet against democracy.* New York, NY: The New Press.

Mosco, V. (2004). *The digital sublime. Myth, power, and cyberspace.* Cambridge, MA: The MIT Press.

Newson, J. A. (2004). Disrupting the "student as consumer" model: The new emancipatory project. *International Relations*, 227–239.

OECD. (2015). *Students, computers and learning. Making the connection.* Paris, France: Author.

Postman, N. (1993). *Technopoly*. New York, NY: Vintage.

Prensky, M. (2001). Digital natives, digital immigrants. *On the Horizon*, 9(5), 1–6.

Serazio, M. (2015). Selling (digital) Millennials: The social construction and technological bias of a consumer generation. *Television & New Media, 16*(7), 599 –615 .

Slack, J. D., & Fejes, F. (Eds.). (1986). *The ideology of the information age*. Norwood, NJ: Ablex.

Smith, J., Skrbis, Z., & Western, M. (2012). Beneath the "digital native" myth. Understanding young Australians' online time use. *Journal of Sociology*, 49(1), 97–118.

Tapscott, D. (1998). *Growing up digital: The rise of the Net generation*. New York, NY: McGraw-Hill.

Traber, M. (Ed.). (1986). *The myth of the information revolution*. London, England: Sage.

ACKNOWLEDGEMENTS

A book project is never quite as simple as you conceive when beginning. There are so many people to thank who have inspired and helped me to process my thoughts on the topic of mobile technologies, global media literacy, and digital citizenry. My sincerest apologies if your name is not listed.

I would like to first thank my family for their continued and unending support in this writing journey. My husband, most especially, who watched me working away in the evenings to make sure that all the details were in place. My Mom, sister, and brother-in-law who have listened quite often to my ideas on the state of media and information literacy, and would in fact keep me up-to-date on global news events.

Vitor Tomé, my wonderful colleague in Portugal, who was so willing to contribute three chapters to this book on a study that continues. It was a true gift for me to have met you in Sweden in 2010 at the World Summit on Media for Children and Youth. I look forward to many more collaborative opportunities. Muito obrigado.

Kat Stewart, my thanks for your thoughtful consideration in writing the foreword to this book. The journey from our original discussions at conferences, co-presentations, and to the work you do every day in the cable industry has had a great impact in the field of media literacy as well as Education. SXSWedu was a journey worth taking with you.

Matteo Stocchetti for your willingness to provide the global perspective on mobile technologies through your preface, which tied the book together so well.

I am most grateful to my colleagues in the field of media literacy, education, and civic engagement: Marieli Rowe, Gayle Bogel, David Zera, Alice Lee, Paul Mihailidis, Ryan Goble, and Melda Yildiz, as their encouragement has helped me keep up with the trends and ideas that are most found in a globalized and mediated society. I have appreciated their thoughtfulness and willingness to provide me with solid thinking while also expanding my consideration of the field. I also want to extend my thanks to my former students at Fairfield University who took this writing journey with me and allowed me to share their thinking.

Thank you to Adrienne Sharpe for being my local editor of the work presented here. I appreciated your questions, suggestions, and corrections. They helped to make this book possible.

Finally, my thanks to Chris Myers, the former editor of this project at Peter Lang, and Sophie Appel for your skills and creativity.

INTRODUCTION

> A society which is mobile, which is full of channels for the distribution of a change occurring anywhere, must see to it that its members are educated to personal initiative and adaptability.
>
> —DEWEY, 1916

Mobile digital technologies have become the frame of reference in education policy and curriculum with many prominent educators discussing the tool as the gateway to knowledge. Yet the idea of literacy and learning has been lost within this frame. Discourse around understanding how to use technology tools is not enough without a context for its use. Mobile tools have proliferated in our global society. There is no one device that is designated as most popular, as, depending on cost, one has superseded the other in terms of usability, likability, and the budgets of those who would purchase them, whether they be businesses or schools. What has been most important is the compact nature of these tools along with the power of what can be accomplished through the applications or programs contained within them. Mobility of technology has contributed to the management of a different type of education system and workforce, making it certainly faster, but even more capable of governing personal and professional articles while also making social interactions simplified.

As part of this acknowledgment of a new mobile capacity, media in all its forms has found a new voice and venue for dispersing messages. For messages that are transmitted via the new vehicle or format that are later interpreted or reinterpreted, and then transmitted back through social networks, open source software, and traditional media, there must be a literacy and learning component. Further, when children, tweens, or teens become the producers of the media messages who regurgitate without consideration of the perception or accuracy of the message, then policy has been misguided by their focus in the wrong direction. The answer for education and learning, then, must focus on critically teaching about digital media literacy and bridging the gap of learning that has been pushed back by assumptions that because a child has a technology tool in hand they are able to understand its implicit learning capabilities.

In the past decade we have seen a substantial growth in the number of mobile devices and applications that our teens and tweens are using. This change has created a disconnect between what is assumed to be learning and engagement versus passive usage. On any given day, you can walk into a middle or high school in the United States and see students who are plugged in at all times. They are visually aware of their friends and maybe even of their teachers, but their focus is on what they are listening to. In that stage, their bodies are active, but their minds are tuned out, or tuned in only to what is coming through their ear buds. In a less active capacity, students with mobile tools are still listening to music, but are now watching the videos. While engaging with their tools they are using social networks to explore posting, primarily on places such as Instagram or Snapchat. They are texting most frequently with friends. Their usage of these devices in this way has proven to be a disruption to learning. Teachers even refer to these devices as "the great distraction," and they are an interruption to the continuum of learning when their instructional voices are eliminated in the process by virtue of the ear bud. Mobile devices have potential, but without governance or plausible usability they can, instead, detract from learning and diminish student/teacher or teacher/student communication, as well as students' communication with the world around them.

The few students who are actively participating in a creative mode are not as visible as those who are engaged in the aforementioned activities. Common Sense Media (2015) recently put out a study that followed a similar model to the 2010 Kaiser Family Foundation, but it broke down the media usage of tweens and teens by the way they used and interacted with their screens. The screens were divided from mobile to other formats. They looked at: mobile gaming, social networking, reading, video gaming, heavy users, and light users. The overall conclusion was that the teen population of heavy users averaged more than 13 hours a day with screen media. As much as 9 hours of that time could be devoted to social networking. More importantly, the numbers and data collected demonstrate that teens today watch TV and listen to music as a primary function. Consumption, not creation, is the end goal of the use of these mobile tools. In fact, the creative potential was lost when entertainment was more favorably valued.

While those numbers alone are incredibly high and they are indeed astounding, they are not surprising, especially not to an educator who on a daily basis fights for their time to teach students in environments that allow mobile technology—in particular, smartphones—to be widely used on a school campus. The struggle with learning while having the availability of these tools is real and occurring every day. We saw a glimpse of this struggle when Mayor Bill de Blasio ended a 10-year-old policy in New York City public schools that had banned cell phone usage, and now openly permits them. Administrators indicated it was "partly as a matter of basic convenience—parents often need to get in touch with their kids, and vice versa—but also as an issue of socioeconomic fairness" (Fertig, 2015; Singal, 2015). Perhaps this was the greater point, that there was a socioeconomic discrepancy, but it didn't solve the problem of how to incorporate mobile phones into the school curriculum so that good learning was taking place along with having an entertainment tool available.

Entertainment and learning, this is in part where the clash of cultures starts to happen, especially in education. Neither has been delineated well over the past 10 years and the current research doesn't demonstrate a positive trend. In a paper entitled, "Technology, Distraction & Student Performance," Louis-Philippe Beland and Richard Murphy (2015) from the London School of Economics discussed that the banning of cell phones for learning is doing the exact opposite. Their study was in part triggered by what had taken place in the New York City public schools, but also by the mere fact that mobile technologies are on the increase and that, ultimately, they would be a part of the learning equation now or in the future. In their study results they stated,

> We find that following a ban on phone use, student test scores improve by 6.41% of a standard deviation. Our results indicate that there are no significant gains in student performance if a ban is not widely complied with. Furthermore, this effect is driven by the most disadvantaged and underachieving pupils. Students in the lowest quartile

> of prior achievement gain 14.23% of a standard deviation, whilst, students in the top quartile are neither positively nor negatively affected by a phone ban. The results suggest that low-achieving students are more likely to be distracted by the presence of mobile phones, while high achievers can focus in the classroom regardless of the mobile phone policy. This also implies that any negative externalities from phone use do not impact on the high achieving students. Schools could significantly reduce the education achievement gap by prohibiting mobile phone use in schools, and so by allowing phones in schools, New York may unintentionally increase the inequalities of outcomes.

Clearly, the economically disadvantaged student and low-achieving student would then be suffering with the distraction of mobile technologies. This is just the beginning when it comes to looking at the issue. More recently, state-mandated test scores, in the United States, revealed that the use of mobile tools such as iPads or Chromebooks did not help to increase students' scores, but instead may have potentially been detrimental. Scores from students who participated in the testing were lower (Herold, 2016). This outcome suggests that perhaps literacy learning must also apply to the tool itself, such as how to read, write, process, and think with a mobile device.

These literacy processes have a definitive implication in the realm of teaching and learning, especially when we know from substantial data collected that the majority of students are not using this tool for anything but amusement or entertainment.

The bigger problem is "the idea" put forth by many policymakers that an iPad, a tablet, a Chromebook, a SMARTboard, or other such tool is going to "save education." This impression has been in place too long and continues to shape what drives curriculum and standardization in the classroom. The reality is that tools for technology are just tools. They don't produce thinking and they don't manufacture materials or ideas on their own. Technology tools need individuals who can make them work. The "work" is what can be created, researched, produced, and made through applications, programs, coding, and so much more. Mobile technologies have provided more opportunities for this type of work to be done without question. The dilemma is, how does this happen when assumptions are made about the users who are, in this case, students who represent the K–12 spectrum?

Bridging thinking and producing with technology is the direction in which this book will be referencing digital media literacy. The concept is not so foreign, but its application takes a student from the most basic understanding to higher level processes, the idea being that through using a mobile device, opportunities for growing in knowledge are possible, can be developed further, and can be repurposed for a greater end—that end being whatever is the motivation for learning and for the individual learner.

How, then, does media literacy and mobile learning fit together? Media literacy education provides for an active voice in the student who is most involved with digital technologies. It also creates a venue for teacher and student to be exchanging ideas using various platforms. For an educator, the mobile tools have the potential for opening doorways for students to a bigger world than that framed by their social networks or their gaming communities. This concept is where the real work of education lies and where this book provides possibilities for the potential theories and discourse.

Mobile learning is here. It is part of the everyday lives of this century of learners. Understanding how that is interpreted in the classroom is vital to its development in mainstream education. Consideration must be given to varied viewpoints. Is it just that access points are given and users can have a doorway? This would be in the realm of information literacy, which is certainly important, and a worthwhile endeavor in any classroom. Is it just for creation? Is that enough? How about that "pleasure principle" that David Considine (2002), an early media literacy pioneer, subscribed to? Is that important to the growing consciousness of how mobile tools have implementation value? The pleasure principle is as basic as it sounds. Each of us as individuals has favorites; we like and dislike many different elements of media and, therefore, make media choices based on our precepts. For our students, the same is true, even if, as educators, we don't necessarily "get it." The point becomes less about "getting it," as it is about understanding why it makes more sense to them and less to us. Instead, it provides an opportunity for other ideas to be introduced and discussed, thereby making the exchange of ideas an important part of the process (De Abreu, 2008). These are the questions and the struggles that are faced by educators globally. The obvious answer is that it must be more.

Access is the most obvious and the most understandable struggle to those who fund educational institutions and academics. It is one area that policymakers can sink their teeth into because it is concrete. Funding for material to make Wi-Fi faster, to build a facility that enables information to be easily attainable, to maintain or upgrade servers, these are all functional items, and important ones, as well, in order for schools to provide working devices to students. However, funding access points does not correlate with education. The funding aspect sometimes tends to lose the educational foundation because providers of monies expect it to happen naturally the minute a system is in place. Unfortunately, in education, one may not always correlate with the other, especially when access points are related more to the functionality of the building, the infrastructure.

The real work happens when all of the material items are in place and teachers can have regular working mobile technologies that have a foundational base in the curricular instruction that is occurring daily. By the same token, it does not mean that mobile technologies supplant already good, functional practice. It doesn't mean because you have access and a mobile tool that it is the only item used. It also should not mean that traditional methods of learning aren't important or viable. This point is perhaps where the biggest mistake is made in education today, where the use of new technology voids already good and practiced educational methods.

The construct of learning is still, to a great degree, to produce transferrable learning, which will be discussed further in Chapter 2 of this book. It is really to gain a place where deeper learning can occur with and through mobile technologies, so that it is applicable in whatever new situation, new digital device, or new teaching/learning method comes into play. It is where anytime/anywhere learning is real and where media literacy education is evident because of the critical thinking skills that are in place. As Tessa Jolls, from the Center for Media Literacy, stated in an interview with Henry Jenkins (2014), the Provost Professor of Communication, Journalism, and Cinematic Arts at the University of Southern California,

> Media literacy education—whether it is high tech or low tech—primarily concerns itself with teaching and learning the conceptual underpinnings beneath contextualizing, acquiring and applying content knowledge. Learners gain content knowledge through using their media literacy skills—and these skills are applicable to any content anytime, anywhere on a lifelong basis. Sometimes this process has little or nothing to do with technology, although I will note that access to technology in the U.S. is widespread: in our experience at CML, in the poorest communities in the U.S., cellphones and applications like video games proliferate, but these technologies are frequently barred in the classroom. This changed education paradigm is a radical shift in cultural and education systems where formal learning worldwide has traditionally been confined to content silos whose subject matter is warehoused in physical textbooks and dumped into students' heads. Since these traditions have dominated since Gutenberg's invention of the press, they are rooted deeply in our culture. "Mastery" is no longer the goal for education; constant improvement on a continuum of learning is what we are seeking, while recognizing that some will inevitably be more skilled than others in various domains.

Thus, the anytime/anywhere concept should be a big part of how mobile technologies learning is incorporated into the educational classroom. But, as Jolls (Jenkins, 2014) stated, education still teaches in the "mastery" mode and is not necessarily looking at learning as a process for moving individuals forward so that they can have a more representative look at their daily lives.

The book undertakes many of the aforementioned subjects in two parts. Part I looks at the link between media literacy and mobile technologies and then explores the mainstream topics encountered along the way with the use of mobile technologies. There are four chapters to begin the exploration of the interconnection between media literacy education and mobile technologies:

- Chapter 1 *Connecting Mobile Learning and Digital Media Literacy*. This chapter opens the discussion on how media literacy education is the better theoretical framework for the development of mobile learning in the classroom.
- Chapter 2 *Technology as a Transliteracy: Creativity and Learning* explores literacy concepts and how they are interrelated. Mobile learning exists as another entity within this construct. This chapter will delve into the role of information literacy, the value of creativity in m-learning and its potential as reflected through various applications and solutions in teaching and learning
- Chapter 3 *Participatory Culture, Civic Engagement, and Equal Access in Practice*. Participating in the world happens naturally through mobile technologies, but how does that produce civic engagement, if at all? At the same, where is the equal access for learning, or even in conceptualization? This chapter will evaluate how these three concepts connect and work together through teaching mobile tools.
- Chapter 4 *Privacy, Student Data—Knowledge as Empowerment*. A look to the future with m-learning will extend the thinking of digital media literacy into the importance of understanding data privacy while managing content with m-learning.

Part II of the book is entitled *Global Perspectives on Mobile Technologies: Online Social Networks-Practices and Perception of Youngsters (9–16), Their Teachers and Parents*. This section highlights how mobile learning has allowed for connections to be made in a more universal way. Through the look at a study conducted in Portugal with Vitor Tomé, the following chapters investigate how mobile technologies have shifted educational methodologies and applications in teaching practice.

- Chapter 5 *Uses and Practices*. This chapter focuses on online social networks use by students (9–16), teachers, and parents, as well as on the most common practices they develop online, namely news consumption, communication with others, posting and sharing habits, and online relationship among them.
- Chapter 6 *Risks and Opportunities*. Teachers and parents are more concerned about the risks than the opportunities of students in social media, mainly because the mainstream media are focused on risky situations. The connection between online and offline risks as well as between risks and opportunities is discussed in this chapter based on the perceptions coming from the field (students, teachers, and parents).
- Chapter 7 *Learning Perceptions: Perspectives From Students, Teachers, and Parents*. Teachers, students, and parents agree that social media have pedagogical potential but they do not see the formal and informal contexts like two sides of a coin, which would be the learning. This chapter focuses both on the potential learning possibilities available through online social networks while also examining the reasons why online social networks are barely used in the classroom.

Lastly, the book has a concluding chapter which brings together both sections of the book, and includes three items in the Appendix section for further exploration: The United States Department of Education National Education Plan Conclusions and Recommendations; Student Privacy Pledge Signers; and Student Data Privacy, Accessibility, and Transparency Act (MODEL LEGISLATION), which has been used nationally.

The expectation in writing this book is that the subject matter provides an exploratory discourse as to the value of mobile technologies through digital media literacy, especially at a time when much of what has changed is the dynamics of classroom learning. Further, this book will look at how technology is shaping thinking without considering the implications of digital media literacy.

References

Beland, L. P., & Murphy, R. (2015, May). *Ill communication: Technology, distraction & student performance*. Retrieved from Centre for Economic Performance, London School of Economics and Political Science website: http://cep.lse.ac.uk/pubs/download/dp1350.pdf

Common Sense. (2015). *The common sense census: Media use by tweens and teens*. Retrieved from Common Sense website: http://static1.1.sqspcdn.com/static/f/1083077/26645197/1446492628567/CSM_TeenTween_MediaCensus_FinalWebVersion_1.pdf?token=swYf74HMaDsU4X5rXZnRIawdhoU%3D

Considine, D. (2002, October). Putting the ME in media literacy. *Middle Ground: The Magazine of Middle Level Education*, 6, 15–21.

De Abreu, B. (2008). *The implementation of media literacy in the K–12 public education classroom: Three case studies* (Doctoral dissertation). Storrs, CT: University of Connecticut.

Dewey, J. (1916). *Democracy and education: An introduction to the philosophy of education*. New York, NY: Macmillan.

Fertig, B. (2015, March 2). Life after the ban: Schools roll out the new cell phone rules. Retrieved from http://www.wnyc.org/story/cell-phone-ban-ends-schools-explain-new-rules/

Herold, B. (2016, February 3). PARCC scores lower for students who took exams on computers. *Education Week*. Retrieved: http://www.edweek.org/ew/articles/2016/02/03/parcc-scores-lower-on-computer.html.

Jenkins, H. (2014, September 15). The value of media literacy education in the 21st century: A conversation with Tessa Jolls (Part Two) [Blog post]. Retrieved from http://henryjenkins.org/2014/09/the-value-of-media-literacy-education-in-the-21st-century-a-conversation-with-tessa-jolls-part-two.html#sthash.SQaRZEvI.dpuf

Singal, J. (2015, May 21). New research suggests de Blasio's unbanning of cell phones in NYC schools may have been a bad idea. Retrieved from http://nymag.com/scienceofus/2015/05/downside-of-allowing-cell-phones-in-schools.html

PART I

LINKING MOBILE TECHNOLOGIES AND DIGITAL MEDIA LITERACY

Part I of this book provides the foundation and argument for digital media literacy through the use of mobile platforms. Each of the chapters represents a collection of thinking that sets the ground work through the utilization of media literacy education, while at the same time exploring more fully mobile technologies as a mediated environment. Across various contexts and with a critical consideration of the educational and practical use of mobility, this section of the book sets the stage for valuing and examining the context for which media literacy is fundamental to an area of technology in constant development.

· 1 ·

CONNECTING MOBILE LEARNING AND DIGITAL MEDIA LITERACY

> Mobile learning is a radical change from the traditional education model. It will transform your school system in ways far beyond simply allowing mobile devices into your classroom. The implementation process will be unique for each school system because at its core this transformation is all about changing the culture.
>
> —SCHAD, 2014

Introduction

Tablets? iPads? Laptops? iPhones? Desktops? Which would you choose? The question faced by educators today is an overwhelming possibility of technology tools, but the discussion of literacy is limited. Whether it is reading and writing literacy in the form of digital language or the interpretation of messages through the use of media literacy, critical thinking is limited or non-existent in many education systems when digital technologies, in particular mobile technologies, have become the conversation of choice.

At issue is the question of what is the purpose of technology and how is it driven in the e- classroom? Extending the discussion further is the concept of "technology" literacy, which has some very unique perspectives depending on whether you consider the technology a form of literacy or the learning that is associated with technology the actual literacy. All of these questions and considerations suggest that there is a change in learning in schools. It is more than just what we believe the current educational platform is today, but also where we are going in the future.

Digital Generation

The way new digital platforms are talked about in schools by administrators and even parents suggests an assumption that students are already digitally literate or digitally savvy. It's a commonly held belief that students learn by osmosis when they pick up the new technology. After all, they are the digital generation and the use of these devices happens to them naturally. They know this stuff, right? Wrong. Recently, while sitting through a keynote at a conference, the presenter pulled up a video of his baby touching an iPad and pushing on the screen. He declared that his child was already digitally savvy as the baby knew where to touch in order to get the tool to work. Audience members made up of educators and parents were astounded by this man's contention. This assessment was alarming, if not erroneous, for many reasons, including the most basic—early educators would say that this is a normal process for childhood development and it has nothing to do with real learning or understanding. It most certainly does not prove that the infant is digitally literate. Sadly, these assumptions from speakers of all levels can often be heard in technology conferences, and then later repeated by policymakers. Even parents will make these assertions because they do not feel that they are as knowledgeable about technology tools as their children. Oftentimes, these comments are misconceived and confused with the idea that today's generation has lived with technology or acquired and assimilated it into their lives since birth in a meaningful, productive way. This generation of students may be more adept at using a device, but that does not equal literacy of any kind.

Many educators who work with students on different technological platforms would agree that many students are digitally illiterate—and that is in its simplest form. In fact, many students struggle with the basic elements of how to access a program or analyze the value of information on a website or database or even how to "cut" and "paste." On the other hand, these same students are very adept on social networks and gaming sites.

> Just as technology is playing an increasing role in culture generally, so too does it play a growing role in the lives of children. Children and young people, then, are actively

> manipulating digital media to participate in social and cultural life outside of the school and making and sharing media has become increasingly important in the way that young people communicate with each other. (Hague & Payton, 2010)

The ability to maneuver around sites is a skill and it does lend itself to learning, but it is a small piece of the puzzle. To be critical and to develop the ability to think through the process of conceptualizing an idea and putting it into action requires much more reflective thinking.

Preparing students to be literate users of technology and to also make discerning choices when involved as content creators is very much needed in education and instruction. The challenge is for teachers to seek out these opportunities within their content areas and the curriculum. In order to understand how pervasive these new tools of technology and formats for networking are to teens, it is important to comprehend children's and young adults' technological behaviors:

- Nearly two-thirds of Americans are now smartphone owners, and for many these devices are a key entry point to the online world (Smith, 2015).
- Seven in ten (71%) children in the UK aged 5–15 had access to a tablet at home by the end of 2014, up from just over half (51%) in 2013, and many kids do not even have to share a tablet with their parents. One in three children (34%) aged 5–15 have their own device – up from one in five (19%) in 2013 (Ofcom, 2015).
- 76% of teens use Instagram, while Facebook has lost the majority of its teen representation (Dewey, 2014).
- The number of US tablet users is forecast to exceed 176 million in 2018 (Statista, 2016).
- Three-out-of-five young adults believe that mobile devices will somehow vaguely rule every aspect of the future (Meeker, 2015).
- Over half of children 3–4 and three quarters of 12–15s use a tablet in 2015 in the UK (Ofcom, 2015).
- The number of children who have their own tablet has also increased to one in seven 3–4s (15%, up from 11% in 2014) and two in five 5–15s (40%, up from 34% in 2014) (Ofcom, 2015).
- Global sales of wearable devices will top 10 million this year, up 32.8 %, which will impact education and the mobile movement (Bolkan, 2016).

As these numbers are continuously increasing, educators must begin to consider how these student content creators and networkers exist in the classroom. This requires a reinforcement of understanding, analysis, and knowledge from the lead educators in many schools, but it does not mean that the technology tool itself will "save" education.

Policy

At the policy level, presumptive statements about a learner's technological ability and knowledge place educators and students at a disadvantage. The dialogue in those circles becomes less about offering learning opportunities for teachers, which would translate to growing their students' technological capacity, and more about buying into more technological tools and platforms along with a focus on the infrastructure. While these elements are important, the lack of attention given to literacy skills as related to digital technologies is a cause for concern. As more and more money is invested in certain types of tools, it also changes the way in which we interact with the different types of tools.

When the iPad was unveiled to schools it was highlighted as the tool that would "save" education. This language puts a high price on the ideology of education, which is based on more than just the fact that schools educate students by way of a device, but through researched, pedagogical curriculum and instruction. The tool never addressed the underlying problems faced in schools such as poverty, social inequity, violence, etc. As Larry Cuban from Stanford University stated in an interview with *The New York Times*, "There is very little evidence that kids learn more, faster or better by using these machines; iPads are marvelous tools to engage kids, but then the novelty wears off and you get into hard-core issues of teaching and learning" (Hu, 2011). Few followed this perceptive statement in the early stages of the inception of the iPad in schools.

The issue of technology in schools has been ongoing since the 1990s when bringing the Internet into the classroom was part of the Clinton/Gore campaign and later the Gore/Lieberman campaign. By politicizing this piece of education, there has been a rush to compete for a benefactor either through federal or state monies or private donors who provided the initial and much needed funds to increase the offerings of technology tools. Unfortunately, by making educational technology a political issue the needs of individual learners are oftentimes ignored or missed. As Stocchetti (2014) indicated, "The idea that education is a fundamentally political process construes education as a process in which a number of participants compete for a variety of goals depending on the nature of the issues at stake" (p. 20). The issue becomes about "haves" and "have nots," with the addition of high-stakes testing in the mix creating more pressure to gain results, as is evident in the United States.

Recently there has been an increase in technology acquisitions for schools with the Google Chromebook. The Chromebook operates by using Wi-Fi capability and also the cloud; however, without Wi-Fi availability, the Chromebook has limited use. Yet the Chromebook is a cheaper tool that is economical for schools to consider for purchase, and schools have done so, but not because of digital equity or to increase students' technological knowledge, but primarily for testing purposes. In administrative circles, the discussion about Chromebooks has been about how they are helping students with their learning, *AND* because they are an inexpensive tool they will, therefore, help to equalize education. When did a technology tool actually do that? While having tools such as these in the classroom certainly helps with learning and with bringing the world a bit closer to the students, the infrastructure must be in place to support these tools, and teacher professional development must be in place. Further, most studies that look at equalizing education and promoting learning are looking at the wider and holistic connection to learning and students—the family and economics. Moreover, the data in the United States demonstrates definitively that "well trained teachers are the greatest school-related contribution to academic success" (Taylor, 2013). However, even bigger than this factor are the things that happen outside of school such as poverty and violence (Rotella, 2013). Those are the more prominent factors for why students are not engaged in the classroom or invested in learning.

Literacy

When the term "literacy" is used it is oftentimes confused with the precepts of print texts or even limitations that don't extend to digital tools. Literacy in its most basic form is defined as the ability to read and write (Alvermann & Hagood, 2000; Buckingham, 2003). This is further discussed in the *NMC Horizon Report* (2014) when looking at the challenges for higher education in adequately supporting and defining digital literacy.

> Digital literacy is an often-debated topic because it encompasses a number of skills that differ between teachers and learners. Failure to redefine literacy in modern times makes it difficult to formulate policies that address low digital fluency of learning facilitators and allow K–12, higher education, community colleges, museums, libraries, and workplaces to take full advantage of the digital tolls and learning resources. (Johnson, Adams Becker, Estrada, and Freeman 2014, p. 2).

Can we not extend those two thoughts to the digital technologies frequented by students who work with them? It is this extension of thought that considers the value of media literacy instruction.

Media literacy requires that we look beyond traditional texts, thereby validating ways in which media play an enormous role in our students' lives. Media literacy promotes critical thinking beyond just the traditional literacies. Potter (2005) suggested that active literacy is the key component and presented the greater thought on the purpose of media literacy within this context:

> Media literacy is a set of perspectives that we actively use to expose ourselves to the media to interpret the meaning of the messages we encounter. We build our perspectives from knowledge structures. To build our structures, we need tools and raw material. These tools are our skills. The raw materials are information from the media and from the real world. Active use means that we are aware of the messages and consciously interacting with them. (p. 22)

Merging an understanding of the media with the actual mediated environment in the lives of individuals who are participating in the media experience is a primary goal of media literacy education (De Abreu & Mihailidis, 2014). In the cycle of learning, they can be on opposite sides of the spectrum or they can share in the knowledge of media. In most cases, no two people will experience the same media environment in exactly the same way. Understanding this perspective enables educators to validate and welcome the different ideas that are presented through the various mediums. Students must be taught to become self-reflective and even critical of the nature of the media for there to be analysis and evaluative thought. As Mihailidis and Fromm (2014) stated, "In today's digital culture, media literacy must necessarily incorporate critical navigation and expression as parallel to analysis and evaluation" (p. 101).

Media Literacy Education

In a contextual understanding of digital media literacy education, it should still be inclusive of the ideas of technology by offering a repository for critical thinking. Media literacy in a digital age can offer opportunities in education whereby teaching and learning can be thoughtful, perhaps critical, but carried across whatever technology platform exists today or in the future. As such, much like traditional literacies that allow for students to have opinions, likes and dislikes, the conversations in the classroom must be open to a dialogue that is respectful, while growing and pushing students' thinking in various directions. Those directions will shift learning where the student is the voice and the leader in the classroom and the educator is the facilitator of the learning. Technology offers an opportunity to extend voice, and teaching students to understand how that voice can be powerful extends far beyond the tool itself.

Media literacy education in the era of digital technology should be a "navigation tool for educators to discuss, challenge, critique, and understand how media, both traditional and new, has impact on students' beliefs, thoughts, behaviors, and more" (De Abreu, 2014, p. 196). And, in a world where these students will be forevermore encompassed in their digital environments, there is an opportunity through media literacy education for creating empathy, addressing issues while being mindful of various points of view, and ultimately creating a global digital citizen who can reach beyond the scope of their medium of choice. As David Buckingham (2010) stated,

> Media literacy, it seems, is a skill or a form of competency; but it is also about critical thinking, and about cultural dispositions or tastes. It is about old media and new media, about books and mobile phones. It is for young and old, for teachers and parents, for people who work in the media industries and for NGOs. It happens in schools and in homes, and indeed in the media themselves. It is an initiative coming from the top down, but also from the bottom up. In these kinds of texts, media literacy is also often aligned with other contemporary "buzzwords" in educational and social policy. It is about creativity, citizenship, empowerment, inclusion, personalization, innovation, critical thinking...and the list goes on. (p. 5)

A student's voice is amplified through digital platforms; therefore, if educators are expected to empower students and create citizens of the world, the value and importance of media literacy education cannot go unstated or untaught.

Mobile Technology

Mobile technology is a large part of mainstream education worldwide. With the advancement of smartphones and other mobile tools, there are candid conversations going on in educational circles as to how this can be repurposed as a tool in the classroom. It already attracts students and encourages the text language development, but it could also be much more. The fact that it is available anytime and anywhere also makes the use of mobiles much more of an enhancement as a means of instructional development. Mobile technologies offer flexibility within the confines of the classroom. This flexibility also exists in the form of communicating between teacher and students, student and teacher, and student to student. As part of this consideration, researchers are looking at the potential for learning processes, whether informal or formal, which are taking place through online chat or instant messaging. As David Buckingham (2007) stated,

> online chat and instant messaging require very specific skills in language and interpersonal communication. Young people have to learn to "read" subtle nuances, often on the basis of minimal cues. They have to learn the rules and etiquette of online communication and to shift quickly between genres or language registers.

The idea of understanding and enhancing language and communication is fundamental to the development of mobile learning.

Since the early part of this decade, a transformation has taken place with the types of ways in which people are communicating online, and specifically with social media. No longer are people using their standalone computers or laptops; instead, their computing has shifted to the mobile device, whether it is in the form of a tablet, an iPhone, or some new, innovative tool, thus enabling them to engage in their social networks more readily and certainly faster. Communication, which was sometimes delayed and compromised, has become wider and greater, while the mode of exchange has become smaller and faster. As a tool for education, mobile technologies and e-learning are in a place of transition and growing.

In the United States, cell phones and tablets were not at their prime until the latter part of 2010, which tied into the release of Apple's iPad that year and the later upgrades to the iPhone series .Yet, the first iPad, while exciting in the technological world, was not quite ready for the educational classroom. When the iPad 2 was introduced, which was crisper, faster, and accessible via all cell phone companies in the United States, the phenomenon took hold. Sales of this product skyrocketed, making Apple stock one of the most expensive to buy (Sperling, 2010). However, educational sales of the iPad are beginning to decline as the realization has taken hold that the tablet alone may not be enough to cultivate learning in the classroom.

Schooling, along with mobile devices, can no longer be considered "one size fits all." With the wide range of mobile technologies as well as other innovations taking place, teaching and learning must continue to transform in order for knowledge to be gained. New approaches and strategies must be considered as the amount of technologies that are influencing learners' lives is continuously evolving. In a study by Common Sense (2012), 90% of respondents from ages 13–17 claimed use of any form of social media. Following this study was one by the Pew Internet & American Life Project, which emphasized that the internet and digital technologies were significantly impacting the way in which students conducted research (Purcell et al., 2012).

Technology has always moved faster than education. This is evident to any person who walks into schools globally. Further, our young children and our teenagers are in the middle of the most social time in our history, and they are active partners in this movement that is shaping their identity and their connection to the world. Students have always been noted for learning new technology at a more rapid pace than most adults, but it is significantly apparent now. Their relationships and their form of sharing information depend on the continuous flow of information that they are regularly connected to. Teenagers take what they learn and then bring it to the attention of their friends. Those friends, in turn, pass it on to other friends, and the cycle continues. It is no longer necessary for this to be downloaded via a desktop computer; their newfound learning can be picked up via the simplicity of a smartphone.

Social Networking

Educators who discount the value of social networking tend to fall into two categories: those who are not sure how to or those who are not using it. The greater issue for many teachers is that they do not actually know how to use social network sites, or find the use of mobile technologies, especially the cell phone, in schools to be a bit precarious for education.

> My phone has become an integral part of my life and communication with the world. All this and yet I was one of those people who would have said that cell phones could be a distraction in the classroom. As I have read more about this issue, I found

> myself more open to the idea, so when I came across the article giving so many concrete examples I was hooked. Not to say that I think the technology (cell phones) should be a primary tool, but certainly a supplemental option for those working with a population with personal access. As I was reading example after example I thought about myself as a student and how I would love participating in polls via my phone and engaging in project learning using my phone. The fact that there are so many applications for use with phones in general, but especially smartphones, I can't imagine that their use won't become commonplace as the technology becomes more and more affordable. The devices themselves are already financially viable, it's the data that is still a real issue for impoverished communities. I do think that media literacy yet again becomes paramount here. I loved the digital dossier video as a tool and think that is a must for student awareness. As the statistics presented pointed out, students are using these devices and this technology on a massive scale and it is only growing. Integrating their use into the classroom even a little opens the door to literacy and etiquette. It feels like an opportunity to at least be discussed in all schools and departments. (Educator)
>
> I think there's a fine line between using cell phones for learning, and being distracted by them in class. First and foremost, I will have a "cell phone only with my permission" policy. I know that even as high schoolers, students have responsibilities outside of school that may actually be more important than what's happening outside class that day. I know that's taboo for a teacher to admit, but let's face it, it's the truth. I will have no issue with students reasonably using cell phones during class if they let me know ahead of time that they have an important reason to. This ensures two things: (1) that I know who is using cell phones and when, and (2) that I understand what pressing issues are on my students' minds during class. With those general guidelines, I'm not a stickler for cell phone use. As long as students don't have headphones on and aren't glued to their phones, let them send that important text, they'll concentrate a lot better with it being out of their minds. I refuse to ever take cell phones away from my students, with the slim exception of test taking. Prior to taking tests, I definitely expect students to place their cell phones in a reasonable place within my sight so there are no concerns about cheating. We're not dealing with 10-year-olds here, we're dealing with smart, capable high schoolers who are on their way to becoming independent individuals, and part of that independence today involves the use of their cell phones. Cell phones are undoubtedly useful for class, they can access thousands of applications with thousands of different educational uses and have immediate connection to trillions of pieces of useful information. However, we need to be cautious about how we deal with this in class. (Student Teacher)

As these statements indicate, educators are trying to find ways to negotiate having the cell phone in the classroom as a tool. Besides trying to understand and incorporate the new applications or "apps" into their classrooms, teachers are trying to find the boundaries for which the use of this tool makes sense. This seeming imbalance has created an atmosphere in schools where either the cell phone is banned completely or full rein is given to students, resulting in an uneasy relationship between students and teachers. In a study published in *American Secondary Education*, teachers had grave concerns and "obviously did not want to be complicit in activities that are illegal, unethical, or unfair to others" (Charles, 2012).

Furthermore, educators continue to struggle with the idea of connecting the cell phone to their already full curricula. They understand that their students use it, enjoy it, and spend a great deal of their time involved with it, but they haven't found insight into this part of their students' lives. This is really where the work in education circles worldwide has to begin, as the increase of knowledge would also help breed confidence in working in those areas that are dominated by youth (Buckingham, 2007). This paradigm shift is precisely where there needs to be more explorations as "Newer 'Smartphones' with multiple capabilities and similar handheld mobile devices are now being used deliberately in some places as a less-costly alternative to laptops in programs encouraging students to engage in academic pursuits" (Charles, 2012). The line between classic written language and the syntax of text language is crossing over into the classroom. Yet the potential for learning exists, and tapping into that area opens teachers to new teaching possibilities given the chance. The most current research shows texting doesn't erode literacy skills. Kemp has noted that "as children learn, they go through a period during which they play around with language, figuring out alternate, creative ways of saying things. Researchers have noted kids doing this for years, in notes and letters—way before texting even existed" (Wood, Kemp, & Waldron, 2014).

Whether it is cell phones with their text features, social sites such as Instagram, Snapchat, or Twitter, this generation of youth has created another platform for communicating and learning, which is not frequently used in many schools across the United States. "A social networking site is only effective as an instructional tool if a school district has a plan for using it...a school district needs to look at a social networking site as a part of a full technology plan" (O'Hanlon, 2007).

As of July 2011, 95% of all teens in the United States ages 12–17 are now online and 80% of those online teens use social media sites (Brenner, 2012). Furthering the statistic is the number of students under the imposed age requirement who have also found sites such as Facebook to be entertaining and a valuable form of communicating. *Consumer Reports*, an agency within the United States, has reported that there are over five million children under 10 years of age on Facebook (Barack, 2011). These increasingly large numbers of active, underage or young users represent a growing and pressing need for educators to understand why students delve into these areas, what learning can come from their participation in these areas, and what knowledge is gained and shared.

Youths need skills for working within social networks, for pooling knowledge within a collective intelligence, for negotiating across cultural differences that shape the governing assumptions in different communities, and for reconciling conflicting bits of data to form a coherent picture of the world around them (Jenkins, 2009). Social networks provide a glimpse of a society that exists, but is usually not met by the hometowns of each of the students. The Internet has made global communications both a possibility and a necessity. More importantly, it has made the ease of connecting real and ever-present. The end result is that this generation of students is characterized as being the first generation to be fully integrated into information communication technology (ICT) since birth—a conclusive fact that needs to be considered and that cannot continue to be ignored by schools, administrators, teachers, or parents (Considine, Horton, & Moorman, 2009).

Meaningful use of educational and information technology is what mobile technologies can offer our technologically invested students. Students of this generation are motivated to look differently at the tools available. Even their exploration through sites such as YouTube can provide the educator with some valuable resources that could be used later with other classes. It is not hard to imagine the possibilities when there are 12-year-old app developers and creators of new tools (Suarez, 2011). At the same time, observing and participating in the items that are engaging today's students supplies the educator with insights into the teenage mind as well as likes and dislikes.

Social network tools that allow active engagement fit in well with the participatory culture and creative connection that is described by Jenkins (2009) in his work on *Confronting the Challenges of Participatory Culture: Media Education for the 21st Century*. He stated that 64% of teenagers have produced digital work and they have done it through the use of mobile technology and social networking sites. The technological vehicle changes the classroom from one that is passive to one that is very much involved with lessons and learning. It enables the propagation of the message that learning is a two-way street and sometimes even multidimensional. Furthermore, it brings the world closer together through the global connections that can be made via the use of any of these tools.

The idea of using mobile phones in learning has been widespread in countries all around the world. Uganda, Afghanistan, and some South American countries have been using the mobile phone as opportunities for collaboration and growth, especially because regular internet connections were poor (Morrison, 2012). Several years ago while in attendance at the International School on Digital Transformation in Porto, Portugal, one of the presenters, Scott S. Robinson (2009), from the Universidad Metropolitana, Iztapalapa Mexico City, discussed this idea of telecenters and cybercafes in this country based on a pilot project that was done throughout Latin America. The project was financed by international donors with the idea of these centers of operation beginning in urban areas and then moving into towns and villages. There were many positives to this program, but some negative ones also occurred, including ownership issues, technology malfunctions as simple as not understanding the climate in which these technologies needed to work, and, ultimately, public policy challenges. However, even with all these hardships there was a distinct resonation that mobile technologies had great potential even in some of the poorest countries. There exists an opportunity that provides for world exchange and a connectedness that would not be present if not for their current existence. With the combination of mobility, social media, and the ability to have wireless access points in many locations, more personalized learning opportunities are bound to happen.

Methods for Learning

These types of conversations are evolving and will continue to increase as more educators, policymakers, and individuals internationally begin to consider the possibilities of a connected classroom on a worldwide scale. Furthermore, on a worldwide scale, a shared vision of where collaboration and growth can happen in the area of mobile technologies will be a continued discussion as well as what kinds of learning can be done through the cloud in the future.

There is a poster that has been sent out through the Twittersphere that asks the question, "What do you want Kids to do with technology?" On the left side there is a list of wrong answers that include: make Prezis, start Blogs, create Wordles, publish Animotos, design flipcharts, produce videos, post to Edmodo, use whiteboard, and develop apps. On the right side are the right answers: raise awareness, start conversations, find answers (to their questions), join partners, change minds, make a difference, take action, and drive change. The column on the right is the most powerful; it impacts education, it will empower students, and it has the best chance of increasing student civic engagement and discourse. In the end, mobile technology won't save education; it will help it, but not save it. On the other hand, media literacy education offers the possibility.

References

Alvermann, D. E., & Hagood, M. C. (2000). Critical media literacy: Research, theory, and practice in "New Times." *The Journal of Educational Research*, 93(3), 193–205.

Barack, L. (2011, May 19). More than 5 million Facebook users are 10 and under, study says. *School Library Journal.*

Bolkan, J. (2016, March 21). Report: Wearable shipments to top 10 million units in 2016. *The Journal.* Retrieved from https://thejournal.com/articles/2016/03/21/wearable-shipments-to-top-10-million-units-in-2016.aspx

Brenner, J. (2012, April 27). Teens fact sheet. Retrieved from http://pewinternet.org/Commentary/2012/April/Pew-Internet-Teens.aspx

Buckingham, D. (2003). *Media education: Literacy, learning and contemporary culture.* Cambridge, England: Polity Press.

Buckingham, D. (2007). *Beyond technology: Children's learning in the age of digital culture.* Cambridge, England: Polity Press.

Buckingham, D. (2010). The future of media literacy in the digital age: Some challenges for policy and practice. *Media Education Journal, 47*, 3–10.

Charles, A. S. (2012). Cell phone: Rule-setting, rule-breaking, and relationships in the classroom. *American Secondary Education*, 40(3), 4–16.

Common Sense. (2012). *Social media, social life: How teens view their digital lives.* Retrieved from http://www.commonsensemedia.org/research/social-media-social-life

Considine, D., Horton, J., & Moorman, G. (2009, March). Teaching and reading the Millennial generation through media literacy. *Journal of Adolescent & Adult Literacy*, 52(6), 471–481.

De Abreu, B. (2014). Grasping the complexities of U.S. educational policy and the classroom: How to move media literacy education forward. In B. De Abreu & P. Mihailidis (Eds.), *Media literacy education in action: Theoretical and pedagogical perspectives* (pp. 193–200). New York, NY: Routledge.

De Abreu, B., & Mihailidis, P. (Eds.). (2014). *Media literacy education in action: Theoretical and pedagogical perspectives.* New York, NY: Routledge.

Dewey, C. (2014, October 8). Teens are officially over Facebook. *The Washington Post.* Retrieved from http://www.washingtonpost.com/news/the-intersect/wp/2014/10/08/teens-are-officially-over-facebook/

Hague, C., & Payton, S. (2010). Digital literacy across the curriculum. Retrieved from the National Foundation for Educational Research website: http://www2.futurelab.org.uk/resources/documents/handbooks/digital_literacy.pdf

Hu, W. (2011, January 4). Math that moves: Schools embrace the iPad. *The New York Times*. Retrieved from http://www.nytimes.com/2011/01/05/education/05tablets.html?pagewanted=all&_r=0

Jenkins, H. (October, 2009). *Confronting the challenges of participatory culture: Media education for the 21st century*. Cambridge, MA: MIT Press.

Johnson, L., Adams Becker, S., Estrada, V., and Freeman, A. (2014). NMC Horizon Report: 2014 K-12 Edition. Austin, Texas: The New Media Consortium.

Meeker, M. (2015, May 27). Internet Trends 2015- Code Conference. Kleiner, Perkins, Caufield, & Byers. Retrieved: kpcb.com/InternetTrends.

Mihailidis, P. and Fromm, M.E. (2014). Scaffolding curation: Developing digital competencies in media literacy education. In Stocchetti, M. *Media and education in the digital age: Concepts, assessments, subversions*, 91–104.

Morrison, N. (2012, November 9). The role of technology in global learning. *The Guardian*. Retrieved from http://www.guardian.co.uk/teacher-network/teacher-blog/2012/nov/06/technology-global-learning-teaching

Ofcom. (2015, November). *Children and Parents: Media Use and Attitudes Report*. Retrieved: http://stakeholders.ofcom.org.uk/binaries/research/media-literacy/children-parents-nov-15/childrens_parents_nov2015.pdf

O'Hanlon, C. (2007, August). If you can't beat 'em, join 'em. *The Journal*. Retrieved from http://www.thejournal.com.theprintarticle/?id=21081

Potter, W. J. (2005). Defining media literacy. *Media literacy, 3rd ed.* New York, NY: Sage.

Purcell, K., Rainie, L., Heaps, A., Buchanan, J., Friedrich, L., Jacklin, A.,… Zickuhr, K. (2012, November 1). *How teens do research in the digital world*. Retrieved from Pew Research Center website: http://www.pewinternet.org/2012/11/01/how-teens-do-research-in-the-digital-world/

Robinson, S. R. (2009). *From telecenters to cybercafes*. Porto, Portugal: International School on Digital Transformation.

Rotella, C. (2013). No child left untableted. *The New York Times*. Retrieved from http://www.nytimes.com/2013/09/15/magazine/no-child-left-untableted.html?pagewanted=all&_r=0

Schad, L. (2014). *Bring your own learning: Transform instruction with any device*. Eugene, Oregon: International Society for Technology in Education.

Smith, A. (2015, April 1). U.S. Smartphone Use in 2015. Pew Research Center. Retrieved from http://www.pewinternet.org/2015/04/01/us-smartphone-use-in-2015/.

Sperling, E. (2010, September 13). Rise of the tablet computer. *Forbes*. Retrieved from http://www.forbes.com/2010/09/10/ipad-apple-blackberry-technology-cio-network-tablets.html

Statista. Tablet users in the United States from 2014 to 2019. Statista Inc. Retrieved from https://www.statista.com/aboutus/.

Stocchetti, M. (2014). The politics of educational reform in the digital age: Concepts, assessments, and subversions. In M. Stocchetti (Ed.), *Media education in the digital age* (p. 20). Frankfurt, Germany: Peter Lang.

Suarez, T. (2011). A 12-year-old app developer. *TEDxManhattanBeach*. Retrieved from https://www.ted.com/talks/thomas_suarez_a_12_year_old_app_developer?language=en.

Taylor, J. (2013). Will technology save public education? *The Huffington Post*. Retrieved from http://www.huffingtonpost.com/dr-jim-taylor/will-technology-save-publ_b_3977533.html

Wood, C., Kemp, N., & Waldron, S. (2014), Exploring the longitudinal relationships between the use of grammar in text messaging and performance on grammatical tasks. *British Journal of Developmental Psychology*, *32*, 415–429.

· 2 ·

TECHNOLOGY AS A TRANSLITERACY: CREATIVITY AND LEARNING

> Literacy is more than the ability to read and write. It is the repertoire of knowledge, understanding and skills that enables us all to participate in social, cultural, and political life.
>
> —BAZALGETTE, 2008

Transliteracy is the "ability to read, write and interact across a range of platforms, tools and media from signing and orality through handwriting, print, TV, radio and film, to digital social networks" (Thomas et al., 2007). Pedagogical theories on mobile learning have taken on a variety of forms including as an interface of media, information, and digital learning. This chapter will discuss some of the most prominent theories along with the blending of media literacy and information literacy, thus exemplifying the definition of transliteracy. Simultaneously, the value of creativity in mobile learning, and its potential—as reflected through various applications and solutions—as the underpinning of teaching and learning is explained.

Frameworks for Mobile Learning

Over the last 5 years, various educational groups as well as researchers have come to some conclusions regarding what an established framework for mobile learning should look like. Futurelab, which merged with The National Foundation for Educational Research (UK) in 2011, provided six broad, theory-based categories in 2004. Their perspective looked at the evaluation of mobile learning initiatives:

- Behaviorist—activities that promote learning as a change in learners' observable actions;
- Constructivist—activities in which learners actively construct new ideas or concepts based on both their previous and current knowledge;
- Situated—activities that promote learning within an authentic context and culture;
- Collaborative—activities that promote learning through social interaction;
- Informal and lifelong—activities that support learning outside a dedicated learning environment and formal curriculum;
- Learning and teaching support—activities that assist in the coordination of learners and resources for learning activities. (Naismith, Lonsdale, Vavoula, & Sharples, 2004)

Another model for framing mobile learning is provided by Marguerite Koole (2009) of the University of Saskatchewan, Canada. Her model consists of a three-circle Venn diagram. The items represented in the Venn diagram are Learner aspect, Social aspect, and Device aspect, which are three main points. The interesting piece about this design is that each of these parts can overlap at any given time and learning is produced. Within the three aspects are criteria device usability, interaction technology, and social technology, with the intersection, mobile learning, being the centerpiece for learning. As Koole stated,

> Mobile learning provides enhanced collaboration among learners, access to information, and a deeper contextualization of learning. Hypothetically, effective mobile learning can empower learners by enabling them to better assess and select relevant information, redefine their goals, and reconsider their understanding of concepts within a shifting and growing frame of reference (the information context).

Based on the Koole model, you can see where mobile learning is occurring when there is an interaction between the learner and their devices. As institutions of learning are looking to embrace more mobile technologies or offer them as a tool for learning, understanding the complexity of the learner and the device is important. Also important is understanding how one device may not serve all instructional purposes, meaning that situating a school with one device can be problematic. This consideration is one that is currently being ignored by many schools nationwide who are purchasing Chromebooks as their solitary model for learning and teaching.

Chromebooks are, frankly, an inexpensive tool that can be brought into the classroom to engage in learning, but it is limited by the mere fact that you can only use Google tools and applications. This mobile tool requires that Wi-Fi services be available or provided in order for it to function. In many places, when standardized testing was placed online, Chromebooks were generally purchased for testing purposes, partly because their average cost was approximately $200.00 USD. The tool became an instrument for testing, but not for much else other than basic technology skills such as word processing. While there is accessibility to online resources when schools are Wi-Fi-capable, the Chromebook does not offer much in terms of the interfacing of applications from other companies and their programs. Further, there is limited higher-order thinking when other more progressive and advanced applications are available through a fully capable laptop or other mobile tool that has provisions for an online and offline life. While the basics of technology are important at the elementary and middle school level, there is a shift in knowledge and skills that needs to occur as students move to high school and later to college and/or careers. Those skills can be attained from the variations of programs to the offerings of classes that require more in-depth skills in order to prepare students for the next phase of their lives. This point brings us back to Koole's (2009) checklist of questions for institutions that are interested in adopting mobile devices and implementing mobile learning. Under consideration should be:

- How might the use of mobile devices change the process of interaction between learners, communities, and systems?
- How might learners most effectively use mobile access to other learners, systems, and devices, in order to recognize and evaluate information and processes to achieve their goals?

- How can learners become more independent in navigating through and filtering information?
- How will the roles of teachers and learners change and how do we prepare them for that change? (Koole, 2009)

All of these questions are valid for considering technology now or in the future, whether it be in the educational classroom or in personal use for students and educators.

More theories of mobile learning are focusing on the convergence of mobile technologies and how students are personally taking on the challenges of mobility. In 1999 The U.S. National Research Council looked at the "idea of contemporary practices that enable successful learning" (Sharples, Taylor, & Vavoula, 2005). Applying them to today's learners and mobile technologies has practicality, especially since the idea behind these specifications is that learning and instruction—in this case, along with mobile tools—produce educational effectiveness. Their learning theory suggests the following parameters:

- Learner centered: It builds on the skills and knowledge of students, enabling them to reason from their own experience;
- Knowledge centered: The curriculum is built from a sound foundation of validated knowledge, taught efficiently and with inventive use of concepts and methods;
- Assessment centered: Assessment is matched to the ability of the learners, offering diagnosis and formative guidance that builds on success;
- Community centered: Successful learners form a mutually promotive community, sharing knowledge and supporting less able students. (National Research Council, 1999)

Understanding this approach to how learning can be developed provides an active platform for engagement with mobile technologies. It also brings about the consideration of Bloom's Taxonomy, the updated version. Bloom's Taxonomy considers the impact of technology on this current generation, and presents an effort to meet and match the evolution of change with students in a mobile and ubiquitous world, thereby emphasizing the active voice and following the thinking process that befits mobile culture (De Abreu, 2011).

Transformational Change

As Toffler (1984) stated, "The illiterate of the 21st century will not be those who cannot read and write, but those who cannot learn, unlearn, and relearn." We need to teach our students to be a part of the learning cycle so that they can grow into the community and future they will face directly in a global world. This must also be true for our teachers, as the cycle of learning in education must be a continuous and well-defined process. Instituting strands of technological learning incorporated into the curriculum must be authentic to the students and to the teachers. Too often reforms in education appear contrived and forced to both educator and student. As Sir Ken Robinson (2011) confirmed,

> The fact is that given the challenges we face, education doesn't need to be reformed—it needs to be transformed. The key to this transformation is not to standardize education, but to personalize it, to build achievement on discovering the individual talents of each child, to put students in an environment where they want to learn and where they can naturally discover their true passions. (p. 35)

Technology can foster student passion and it can also change the way a student approaches problems and solutions when learning. Moreover, technology can provide access points to differentiated learning. The tools that are used can be selected or determined based on knowledge, ability, and motivation. As Richard Strong (Strong, Silver, & Perini, 2001) pointed out, "different students need different styles of instruction to acquire learning, different students need different modes of assessment to demonstrate their learning, and each child's interests and talents, concerns, and fascination are the best foundation and motivation to learn" (pp. 65–66). Students can be the directors of their learning and their teachers' partners in the process. Facilitating learning is a step toward the future where classes can be flipped and student voice instrumental in guiding achievement. Our students' worlds are changing and never has that been more evident than by way of these technological advances to date.

Educators must acknowledge and understand fully how much the digital world is impacting the learning that is transpiring in the classroom, simply by the mere fact that mobile technologies are in existence everyday and in almost every possible venue. This form of learning and teaching falls within the auspices of digital media literacy. In order for that to happen, a deeply thought-out consideration for the subject area along with an understanding of students' mediated environments must come together.

Four Tenets of Media Literacy Education

Media literacy is the ability to access, analyze, evaluate, and communicate information in a variety of forms and formats. These terms are ones that most educators will recognize as being a part of the original version of Bloom's Taxonomy that promotes the concept of high-level critical thinking (Anderson & Sosniak, 1994). The later years have promoted a different movement in Bloom's Taxonomy that is more active in language—fitting of the way in which technologies have allowed the user to be involved in the flow of information. No longer is the participant passive, but actively involved through creating, engaging, or delivering information.

Media literacy promotes critical thinking beyond the traditional literacies of reading and writing, including the visual and computer literacies. Potter (2005) suggested that these literacies are the key components that make up the greater thought on what media literacy is:

> Media literacy is a set of perspectives that we actively use to expose ourselves to the media to interpret the meaning of the messages we encounter. We build our perspectives from knowledge structures. To build our structures, we need tools and raw material. These tools are our skills. The raw materials are information from the media and from the real world. Active use means that we are aware of the messages and consciously interacting with them. (p. 22)

Further, the student voice becomes one of the dominant factors in media literacy education, and it also becomes an opportunity to flip the learning in the classroom, another avenue for where mobile learning can take precedence (Buckingham, 2003; Yildiz & De Abreu, 2013). Students can provide a viewpoint on how the media have been a part of their consciousness. A discourse in media can potentially serve to be a transformative component to traditional literacies in the classroom, whereby a significant connection can be made from literature to film to television, and other online environments. This transference of learning is necessary within the current educational paradigm. Teachers are often asked to incorporate critical thinking in instruction. As educators, we ask students to be critically literate in our content subject areas, but very often forget to ask them to be just as literate about the media. Since the nature of media is alluring and students connect so readily, absorbing any and all material, there is no question that there is a need for further understanding (Considine, 2002).

In 1999, David Considine (Considine & Haley, 1999), a professor at Appalachian State University, provided 10 points that delved further into the issue to make the case for media literacy in schools. Media literacy:

1. Redefines literacy based on the information revolution.
2. Is a competency, not a course and therefore interdisciplinary.
3. Is experiential, hands-on and consistent with learning styles.
4. Fosters responsible citizenship in a democratic society.
5. Nurtures group work, cooperative learning and partnership.
6. Is sensitive to stereotyping, bias and multicultural education.
7. Is consistent with the critical thinking skills movement.
8. Is integral to key components of health/wellness education.
9. Has been successful with at-risk students and retention rates.
10. Connects the curriculum of the classroom with the curriculum of the living room.

Increasingly students are swamped with information overload, and they are making decisions and selections based upon what they are told by, or through, the media. As the information is being forced upon them, many of our students are left to think about what they are seeing and hearing without having a chance for some form of discourse. Understandably, real-world issues and popular culture texts can sometimes be difficult to discuss, either because the content is problematic, or because it requires a judgment call by the educator as to its appropriate use in the public education classroom. Yet, students are in need of teachers who can venture into their "play" areas in order to grasp what they are experiencing. Educators can no longer ignore how our teen/young adult audience negotiates meaning from these media environments and, in particular, mobile technologies, which are the biggest focus of this generation's attention.

The current mode of learning is still very much teacher-directed with very little discourse allowed by the student learner. In fact, the direction of education has been stilted by the mere fact that there is a lack of recognition that today's learner is growing in the exact opposite direction than what is prescribed by the current educational model. The growth of technology and new literacies has created this gap between the teacher-learner relationships, but the overall system of schooling has also been to blame. Many of our schools are still entrenched in the top-down teaching model, but our students' world is collaborative and motivated by the idea of sharing. We are in a world in which our student population is the first generation to have lived with technology since birth (Jacobs, 2010). Our students are active viewers of, and engagers with, all popular culture texts in all formats. The role of media literacy in the classroom should not deny that popular culture has prominent importance in the lives of individuals, but instead channel that energy so that it can help to uncover codes and conventions that can be influential, disempowering, and manipulative (Shor, 1999). This theme is further exemplified by the "20 Important Reasons to Study the Media" developed by Chris Worsnop (1989) in Canada:

1. Like history, because the media interpret the past [to show] us what has gone into making us the way we are.
2. Like geography, because the media define for us our own place in the world.
3. Like civics, because the media help us to understand the workings of our immediate world, and our individual places in it.
4. Like literature, because the media are major sources of modern culture and entertainment.
5. Like literature, because the media require us to learn and use critical thinking skills.
6. Like business, because the media are major industries and are inextricably involved in commerce.
7. Like language, because the media help define how we communicate with each other.
8. Like science and technology, because the media help us to learn technology by adopting the leading edge of modern technological innovation.
9. Like family studies, because the media determine much of our cultural diet and weave part of the fabric of our lives.
10. Like environmental studies, because the media are as big a part of our everyday environment as are trees, mountains, rivers, cities and oceans.
11. Like philosophy, because the media interpret our world, its values and ideas to us.
12. Like psychology, because the media helps us (mis)understand ourselves and others.

13. Like science, because the media explain to us how things work.
14. Like industrial arts, because the media are carefully planned, designed and constructed products.
15. Like the arts, because through the media we experience all the arts as no other age has ever done.
16. Like politics, because the media bring us political and ideological messages all the time—yes—all the time.
17. Like rhetoric, because the media use special codes and conventions of their own languages that we need to understand.
18. Like drama, because the media help us understand life by presenting it as larger- than-life, and compel us to think in terms of the audience.
19. Like Everest, because they are there.
20. BECAUSE THE MEDIA GO TO GREAT LENGTHS TO STUDY YOU!

In order for media literacy to be considered critical literacy, all voices must be heard. There must be a participatory nature to the language of thought and learning providing for all student voices and perspectives to be heard. Those requirements are part of the demands and the challenges faced in education (Buckingham, 2003; Jenkins, Purushotma, Clinton, Weigel, & Robinson, 2006). Furthermore, there must be connections to the students' prior knowledge, cultural background, and community in order for critical media literacy to work well in and out of the classroom (Yildiz & De Abreu, 2013). In the end, the role of the future educator will be to help students understand and interpret the media. As Strasburger (quoted in Jolls, 2015) indicated in an interview with the Consortium for Media Literacy,

> Rote memorization should be a thing of the past. What we need to teach is much more about critical thinking, much more about how to decipher all this information that's available on the Internet, some of which may be accurate, some of which may not be accurate—how to deal with tweeting, and texting, and downloading, and we're simply not doing that. I think teachers are making a huge mistake. A lot of schools think if they have a computer lab and they pass out iPads to all the students that they're keeping current, but that's just scratching the surface. Media literacy doesn't start with an R, but it has to be the fourth R.

Students have a wide choice when it comes to media preferences, especially with the number of online platforms that allow for social engagement. They also have a new range of mobile tools to select from, including the mobile phone, the new tablets from a wide range of companies, and, of course, the ever-present iPad in all its shapes and sizes. They can select different genres of music, different artists, or they can mix-and-match through iTunes and other mobile applications. Television offers the same wide variety of selections, as almost every company has a mobile application that can be used as a viewer on mobile technologies. Students photograph and record themselves, placing themselves in locations or in situations that exist and sometimes do not exist. They are open to being freely engaged online and on multiple platforms such as Instagram, Twitter, Facebook, and others. Students can easily place themselves in the media sphere and do so quite readily. Because it is such a critical part of their landscape, the importance of being able to deconstruct what they see, hear, buy, or select makes the teaching of media literacy valuable (Ito et al., 2008).

Transdisciplinary Approach: Integrating State and Common Core Standards

The Common Core State Standards (CCSS) are the United States' version of national standards for education. While the attempt of the Common Core was to provide school districts within all the states with a formula and format for curriculum and instruction, it took on a very negative connotation. High-stakes standardized testing has become a hallmark of modern school reform, and it was introduced with the CCSS. Therefore, instead of an opportunity for growing learning and instruction in the classroom, it became a punitive measure hurting both students and teachers. Many educational software, textbook, supplements, and assessments companies began developing products to fill the schools, whether it was the PARCC (Partnership for Assessment of Readiness for College and Careers) version or the SBA (Standards Based Assessment) version because it was lucrative (Strauss, 2015). Then, educators were told that computer-based assessments were going to be given to students. In taking the test, they could use any computer device, including mobile ones, as long as they had a keyboard. In many places, that eliminated taking the test with the iPad. However, an even bigger issue was that there were not enough devices available to students in every district. There was also no consideration of the fact that schools in the United States or within each state did not have a good or viable infrastructure in place, and therefore would have difficulties with accessing the testing sites simultaneously. These issues continued, and further highlighted the problems with digital equity in schools nationwide. This topic will be further discussed in the next chapter.

While each state in the United States has the right to include or not include the state standards, federal monies were attached to these content standards. States were told that these monies would be withheld if they did not comply and did not test. Providing you with this background information enables you to understand there is a shift in education at the moment, and these standards may go away completely with the next president of the United States. As of right now, most states are already backing out of using them as their format for instruction. States are instead devising their own methods and ways of assessing students at a much lesser cost than what is currently in place.

> In 2012, a report by the Brookings Institution found $669 million in direct annual spending on assessments in 45 states, or $27 per student. When you add administrative costs involved in tests, the total spent on testing rises to $1100 per student. (Kamenetz, 2015)

In the meantime, the Common Core Standards have enabled integration of digital media literacy, mostly because they were not put in as a measurable tool or with much dictate. This change has come to the relief of many teachers who have noticed that there is a divide in the way in which teaching and learning is happening from class to class. No longer are students able to see the relevancy of what they are learning in math and connect it to reading or writing. Science, social studies, and other subject areas have become very contained, because testing has made teaching and learning exist in silos. The idea of the transdisciplinary approach from early childhood and elementary school curriculum has become nonexistent once students get to middle school. This lack of cross-curricular understanding has created a significant break in the way in which students understand learning (Yildiz & De Abreu, 2013).

Information Literacy

In a mobile technology world, the main purpose is for students and teachers to connect, access, analyze, evaluate, and create on digital platforms, while forming communities of learning with shared interests. Twenty-first-century learning requires teachers and students to design a partnership in which the flow of information can exist with the knowledge that technology is ever-changing.

The American Association of School Librarians (AASL) has been seeking to promote these skills through their definition of information literacy, "a transformational process in which the learner needs to find, understand, evaluate, and use information in various forms to create for personal, social, or global purposes" (American Association of School Librarians and Association for Educational Communications and Technology, 1998). AASL broadens the definition through the concept of critical thinking: "Students need to be better observers, appliers, and evaluators of ideas and information....The ability to think critically appears to be cross-disciplinary, training approaches for students can and should be incorporated into many curricular areas." Similar to media literacy, information literacy asks students to question the common perceptions and push the analysis further through investigation, evaluation, and discernment from various platforms, databases, resources, and applications.

Schools in the United States are at a point of transition, whereby they are beginning the shift of standards from CCSS to whatever will be individually designed state-to-state, and to the new Every Student Succeeds Act (ESSA), which replaces No Child Left Behind (NCLB). They are also deciding if learning will remain within brick and mortar, or if we will begin to shift to online learning or a hybrid model. One of the most significant aspects of any curriculum is the realization that digital technologies must be a part of the learning process. For the first time, there is an acknowledgment that educational technologies have flipped the teaching process on its ear and that, as educators, we must respond to this growing change in our school communities. As was stated in a report produced by the National Association of State Boards of Education (NASBE) (2012),

> For educators and policymakers, one of the keys for effectively responding to this generation is remembering that educational technology is both a tool and a game changer. As digital influences expand and their effects on students' lives increases, some of students' fundamental educational experiences require change.

Creativity and Learning

There is a lingering fear among teachers that creativity is gone. That fear exists despite mobile technologies or the idea of mobile learning. In fact, this assumption, which is a general conversation among language arts teachers, has reached a worldwide audience, with a number of influential educators appearing in current educational literature and conferences as keynote speakers, considered leaders in the field. The expressed worry is that we are producing students who will not be able to think creatively or critically in a classroom environment where the message imparted is that the task of teaching is "to the test."

At a recent educator conference, the presenter stated that when looking at the CCSS, "Creativity was left out" (ISTE 2015). Teachers hearing this statement were startled. For many educators, who are seeking a way out of the constraints of the previous 10 years of No Child Left Behind, the CCSS was supposed to generate hope that their curriculum would be diversified and that to a certain extent they would have the freedom to teach what they felt was most needed or best served their student populations. Instead, this comment, among others, left the teachers in that room deflated and concerned that this new era of teaching and learning would be just as constrained as the previous.

The discussion of the lack of creativity is not new. In fact, it is a common fear that is expressed by many educators, business leaders, and those interested in the development of education nationwide. Ken Robinson (2011) reemphasized this point in *Out of Our Minds: Learning to Be Creative*:

> When I talk with business leaders, they complain that education isn't producing the thoughtful, creative, self-confident people they urgently need: people who are literate, numerate, who can analyze information and ideas; who can generate new ideas of their own and help to implement them; who can communicate clearly and work with other people. (p. 15)

Why is this? Why are people who have an interest in education certain that creativity in schools has died? Unfortunately, many states have followed a "state-mandated, cookie-cutter curriculum"—to the extent that teachers did not have to "think" about teaching, but could also script what was needed and what was being dictated to them in order to sufficiently make/pass the test. For years, many teachers who strive to be creative in their own teaching have found that instead, all they are allowed to do is conduct classrooms that bore the students and themselves. As was stated in *Parenting* magazine, "NCLB has proven itself to be a classic case of good idea, bad execution. It's a noble effort that ended up degrading teachers' morale and putting so much frantic focus on testing that real education got left behind" (Rubenstein, 2012). Real education— where the teacher crafts lessons and ideas that ask students to think and then generate more thinking—is what is most needed. Not rote dictation and regurgitation. The achievement gap, which has been politicized for the last decade, has not been narrowed, especially when programs such as Race to the Top and other federal monies depend on more testing and score-keeping. Schools have taken these measures too literally where, instead of worrying about fostering growth and knowledge, they have been treated to data and more data that amounts to very little when we think of the individual about whom we are collecting data…a 5-year-old, a 10-year-old, or a 15-year-old.

When *Education Week* asked in 2012 for people to comment on NCLB ("NCLB: In a Word"), they created a word cloud of the most popular answers. Besides the more obvious terms of "failed" and "flawed," the other words that are glaringly obvious are "stifling creativity." Indeed, students who have been in schools during this timeframe have not had the opportunities to generate original thought, become entrepreneurs, or work at collaboratively creating something that is unique or worthwhile. "Teaching to the test" became mainstream, and very little class time was given to other parts of the curriculum. This change is evident by the number of art, music, and theatre programs that have been cut over that timeframe because they were considered less important than reading or writing:

> Fewer public elementary schools are offering visual arts, dance and drama classes than a decade ago, a decline many attribute to budget cuts and an increased focus on math and reading. The percentage of elementary schools with a visual arts class declined from 87 to 83 percent. In drama, the drop was larger: From 20 percent to 4 percent in the 2009–10 school year. Music at the elementary and secondary school levels remained steady, though there were declines at the nation's poorest schools. (Armario, 2012)

The International Society for Technology in Education (ISTE) has been a huge proponent of digital tools, creativity, and learning. The ISTE NETS-T (National Education Technology Standards) states as part of their standards that teachers need to "facilitate and inspire student learning and creativity" (The International Society for Technology in Education, 2013). This goal is to move beyond just the knowledge of the subject matter; it also extends to innovation in the classroom as well as on digital platforms that foster virtual collaboration and engagement. It is not the only one who sees the practice of using digital tools in conjunction with curricular learning as an avenue for progression of creativity.

Scholars who are interested in the growth of learning and student engagement pinpoint new technologies as a source for where such changes can take place. Sefton-Green (2013), a British scholar, recently wrote, "it is equally impossible to ignore the fact that the last two digital decades have closely entwined ideas about creativity with our emerging understanding of how digital technology works in practice." The evolution of technology has also precipitated moments where the most creative and innovative works can be accomplished. Some of those works are designed by students, and mostly outside of the four walls of the classroom.

Ultimately, serving our students is going to be about crafting new ideas and new ways of engaging them that will not follow the traditional protocols of the "old" education model. No one can say yet whether new standards will help or hurt our students, but the one thing that is clear is that if education continues to be in direct opposition to the natural progression of where student learning exists, then we are in for some difficult years.

Creativity and critical thinking are not polar opposites and have a place in schools and within mobile learning. However, the training that prepares teachers to instruct using this model could be what destroys whatever innovation teachers could foresee, if they are not allowed to extend their students' experiences and learning in the digital mobile classroom.

Concluding Thoughts

The number of digital tools and the influence of the media have never been more powerful than in this day and age. Our students have also changed because of many of these outside forces that have come into the classroom. The shift becomes: how can we focus on teaching and learning on behalf of our students when there is so much else happening in the world? Integrating our technology curriculum is vital and important to the work we do now and in the future, and understanding of mobile learning has priority.

As Howard Rheingold (2011) noted,

> The mindful use of digital media doesn't happen automatically. Thinking about what you are doing and why you are doing it instead of going through the motions is fundamental to the definition of mindful, whether you are deciding to follow someone on Twitter, shutting the lid of your laptop in class, looking up from your Blackberry in a meeting, or consciously deciding which links *not* to click. (p. 1)

We know that technology has changed our world. We know our students are a part of that world. As educators, we know we need to bridge those two pieces together.

References

American Association of School Librarians and Association for Educational Communications and Technology. (1998). *Information power: Building partnerships for learning*. Chicago, IL: ALA.

Anderson, L. W., & Sosniak, L. A. (Eds.). (1994). Bloom's taxonomy: A forty-year Retrospective. In *Ninety-third yearbook of the National Society for the Study of Education, Pt. 2*. Chicago, IL: University of Chicago Press.

Armario, C. (2012, April 2). Report: Arts classes at elementary schools reduced. Retrieved from http://archive.boston.com/news/nation/articles/2012/04/02/report_arts_classes_at_elementary_schools_reduced/

Bazalgette, C. (2008). Transforming literacy. In U. Carlsson (Ed.), *Empowerment through media education: An intercultural dialogue* (pp. 245–250). Goteborg, Sweden: International Clearinghouse on Clearing, Youth and Media, Nordicom, Goteborg University.

Beghetto, R. A., & Kaufman, J. C. (2013, February). Fundamentals of creativity. *Educational Leadership*, *70*(5), 10–15.

Bronson, P., & Merryman, A. (2010, July 10). The creativity crisis. *Newsweek*. Retrieved from http://europe.newsweek.com/creativity-crisis-74665?rm=eu

Buckingham, D. (2003). *Media education literacy, learning and contemporary culture*. Malden, MA: Blackwell.

Conforth, D. (2012, March 16). Frameworks for mobile learning. Retrieved from https://mobilelearninginfokit.pbworks.com/w/page/41753215/Frameworks%20for%20mobile%20learning

Considine, D. (2002, October). Putting the ME in media literacy. *Middle Ground: The Magazine of Middle Level Education*, 6, 15–21.

Considine, D., & Haley, G. (1999). *Visual messages: Integrating imagery into instruction*. New York, NY: Libraries Unlimited. Retrieved from http://www.medialit.org/reading-room/case-media-literacy-todays-schools.

De Abreu, B. (2011). *Media literacy, social networking, and the Web 2.0 environment for the K–12 educator*. New York, NY: Peter Lang.

The International Society for Technology in Education. (2013). ISTE.NETS-T Advancing digital age teaching. Retrieved from website: http://www.iste.org/docs/pdfs/nets-t-standards.pdf?sfvrsn=2

Ito, M., Horst, H., Bitttani, M., boyd, d., Herr-Stephensons, B., Lange, P.,...Robinson, L. (2008, November). *Living and learning with new media: Summary of findings from the digital youth project*. Retrieved from The John D. and Catherine T. MacArthur Foundation website: http://digitalyouth.ischool.berkeley.edu/files/report/digitalyouth-WhitePaper.pdf

ISTE 2015. Connected learning. Connected world. Retrieved from https://www.isteconference.org/2015/.

Jacobs, H. H. (2010). *Curriculum 21: Essential education for a changing world*. New York, NY: ASCD.

Jenkins, H., Purushotma, R., Clinton, K., Weigel, M., & Robinson, A. (2006). *Confronting the challenges of participatory culture: Media education for the 21st century*. Retrieved from http://www.newmedialiteracies.org/wp-content/uploads/pdfs/NMLWhitePaper.pdf

Jolls, T. (2015, November 19). Voices of media literacy: International pioneers speak: Victor C. Strasburger, MD. Retrieved from http://www.medialit.org/reading-room/voices-media-literacy-international-pioneers-speak-victor-c-strasburger-md

Kamenetz, A. (2015). *The test: Why our schools are obsessed with standardized testing—But you don't have to be*. New York, NY: PublicAffairs.

Koole, M. L. (2009). A model for framing mobile learning. Retrieved from http://www.aupress.ca/books/120155/ebook/02_Mohamed_Ally_2009-Article2.pdf

Naismith, L., Lonsdale, P., Vavoula, G., & Sharples, M. (2004). Literature review in mobile technologies and learning. *Futurelab Series* (Report 11). Bristol, England: Futurelab.

National Association of State Boards of Education. (2012, January). Born in another time: Ensuring educational technology meets the needs of students today—and tomorrow. Retrieved from http://www.nasbe.org/study-group/technology-study-group-2012/

National Research Council. (1999). How People Learn: Brain, Mind, Experience, and School. Washington, DC: National Academy Press.

NCLB: In a word. (2012, January 5). *Education Week*. Retrieved from http://www.edweek.org/ew/collections/nclb-10/

Potter, W. J. (2005). Defining media literacy. In *Media literacy, 3rd ed.* (pp. 15–39). Thousand Oaks, CA: Sage.

Rheingold, H. (2011). *Net smart: How to thrive online*. Boston, MA: MIT Press.

Robinson, K. (2011). *Out of our minds*. Sussex, England: Capstone.

Rubenstein, G. (2012). No child left behind: The good and the bad. *Parenting*.

Sefton-Green, J. (2013, September 23). Is there such a thing as digital creativity? Retrieved from http://dmlcentral.net/blog/julian-sefton-green/there-such-thing-digital-creativity

Sharples, M., Taylor, J., & Vavoula, G. (2005). *Towards a theory of mobile learning*. Paper presented at MLEARN 200, Learning Sciences Research Institute. Retrieved from http://www.oecd.org/edu/ceri/38360564.pdf

Shor, I. (1999). What is critical literacy? In I. Shor & C. Pari (Eds.), *Critical literacy in action*. New York, NY: Heinemann Press. Retrieved from http://www.lesley.edu/journal-pedagogy-pluralism-practice/ira-shor/critical-literacy/

Strauss, V. (2015, March 11). Five reasons standardized testing isn't likely to let up. *The Washington Post*. Retrieved from https://www.washingtonpost.com/news/answer-sheet/wp/2015/03/11/five-reasons-standardized-testing-isnt-likely-to-let-up/

Strong, R., Silver, H., & Perini, M. (2001). *Teaching what matters most: Standards and strategies for raising student achievement*. Alexandria, VA: ASCD.

Thomas, S., Joseph, C., Laccetti, J., Mason, B., Mills, S., Perril, S., & Pullinger, K. (2007, December 3). Transliteracy: Crossing divides. *First Monday*, *12*(12). Retrieved from http://journals.uic.edu/ojs/index.php/fm/article/viewArticle/2060/1908

Toffler, A. (1984). *Future shock*. New York, NY: Bantam.

Worsnop, C. (1989). 20 important reasons to study the media. Retrieved from http://www.medialit.org/reading-room/20-important-reasons-study-media

Yildiz, M., & De Abreu, B. (2013). Fostering global literacies among pre-service teachers through innovative transdisciplinary projects. In *Literacy Enrichment and Technology Integration in Pre-Service Teacher Education*. New York, NY: IGI Global.

· 3 ·

PARTICIPATORY CULTURE, CIVIC ENGAGEMENT, AND EQUAL ACCESS IN PRACTICE

> The genius of democracies is seen not only in the great number of new words introduced but even more in the new ideas they express.
>
> —ALEXIS DE TOCQUEVILLE

> We are witnessing a fundamental change in how individuals can interact with their democracy and experience their role as citizens. Ideal citizens need not be seen purely as trying to inform themselves about what others have found, so that they can vote intelligently. They need not be limited to reading the opinions of opinion makers and judging them in private conversations. They are no longer constrained to occupy the role of mere readers, viewers, and listeners. They can be, instead, participants in a conversation.
>
> —BENKLER, 2006

Aaron Sorkin's television series, *The West Wing,* debuted on September 22, 1999, and ended on May 14, 2006. Perhaps one of the most well-written shows about interpersonal relationships, civic engagement, and the world of politics, it has retained an audience while in syndication, and its impact is still discussed years later. *The West Wing* offered a blending of entertainment and education about American politics. The education came through the various characters, whether it was White House communications director Toby Ziegler informing in his caustic manner his frontline person with the media,

White House press secretary C. J. Cregg, or personal aide to President Josiah Bartlett, Charlie Young, challenging his protégé Anthony on his knowledge of the First Amendment or "separation of church and state." This program provided viewers with an understanding of what it was to participate in government, to have civic engagement, whether through peaceful demonstrations or otherwise.

While the show was fictitious in its character build-up, many of the storylines followed some real-world events that were occurring during the time period of the show. At the same time, it was not a completely accurate portrayal of the presidency. However, the show was credited with bringing people to the world of politics and policy: "On more than one occasion, *The West Wing* has been called a weekly national civics lesson that provides viewers with an insider's perspective of the inner-workings and complex praxis of national governance in the White House" (Dumlao, 2005).

This chapter, which covers the spectrum of participatory culture, civic engagement, and the digital divide, exists to exemplify one thing: media literacy and critical thinking are converging in public dialogues provided by mobile technologies and through the Internet.

> Critical thinking is an essential skill for the 21st century students. Not that it was not essential before but in a digitally focused world where there is an exceeding surplus of amateurish and unpolished information, the ability to critically assess, weigh in and value information has an imminent urgency. Through critical thinking, learners will get to sift through the ocean of information provided by the net and be able to decide what is relevant and what is not. Critical thinking embraces several conceptual skills that include synthesizing, assessing, analyzing, applying and evaluating information. Critical thinking allows students to make informed and rational decisions as it forces them to dive deeper and analyze things logically. What would be viewed as a common sensical fact by "surface thinkers" would be considered as a provocative statement that invites much reflection and investigation by critical thinkers. ("Critical Thinking Explained to Students," 2016)

Provocative, and a necessity when looking at how audiences in education circles, whether students or adults, interact or do not interact in forming a path to digital citizenship. This concept also reinforces the idea that "Media literacy is an extremely important factor for active citizenship in today's information society" (Zaccheitti & Vardakas, 2008).

Participatory Culture

A participatory culture is defined as "culture with relatively low barriers to artistic expression and civic engagement, strong support for creating and sharing creations, and some type of informal mentorship whereby experienced participants pass along knowledge to novices" (Jenkins, Purushotma, Clinton, Weigel, & Robinson, 2006). What do we mean by participatory culture when discussing mobile technologies? In essence, our youth participate in using digital formats while creating, dialoguing, and networking—digital formats that enable the creation of mash-ups, playing video games, social networking, podcasting, dialoguing on fan sites, or their involvement in the many different online capacities that are of particular interest to them. They are engaged when they are interacting with sites, especially when they are using them with other people who are also participating online, but in some cases their participation is limited to usage.

At the same time, while Pew studies (out of the Pew Research Center) have demonstrated that more than 50% of students are media producers, the issue is this categorization (Perrin, 2015). Right now it appears as a sweeping generality that does not accurately portray the youth culture and coincides with this idea that they are digital natives and therefore digitally savvy, versus the reality that they just live in a digital world and may not actually be using these tools for anything more than consuming.

The importance of the idea of participatory culture is to understand where youth are at when using technological consumables that have creation possibilities. Are they low-level participatory contenders or high-level participatory contenders? What distinguishes the levels? The difference between a low or high rating would be in the actual participatory task. For example, if a student e-mails, that would be in the low category; a person who blogs would have a higher categorization. A person who creates vlogs or creates applications online, while dialoguing on sites that support the work done, would also receive a higher categorization because there is more development of thought, critical thinking, and skill. A person who dialogues while playing a video game and strategizes new approaches through an online platform would also receive a higher categorization, but just playing without engagement would receive a lesser rating.

Understanding the nature of the involvement in online activities brings us to ask, does a participatory culture actually do anything? That question is what leads to the next action on civic engagement.

Civic Engagement

In the current election cycle, it is obvious that one candidate has attracted the youth vote through the use of social media—Bernie Sanders. In 2008, then-President-elect Barack Obama used Facebook to effectively garner representation from youth. The trend is beginning to take shape that demonstrates that civic engagement is taking place in a variety of ways and through mobile technologies.

The Internet has widened our scope of the world by providing us all with a doorway to increasing amounts of information. Whether the information is in the United States or is coming from other countries, the value of the exchange of information has changed the way in which individuals see the world or interact with it. In a chapter, in my previous work, *Global Media Literacy in a Digital Age* (De Abreu and Yildiz, 2016), it was made evident through research that people in Pakistan and Afghanistan learn more about military actions in their countries by accessing online news sites than through their local newspapers. As Martin & Nakayama (2010) state, "In some ways, the Internet has democratized information, in that more people control and disseminate information than ever before." In spite of this and other governments' attempts to limit their citizens' access to computer-mediated communication, the Internet is providing information, world news, and possibilities for interpersonal communication that were not available previously (De Abreu, 2016).

Facebook, but most especially Twitter, has had a major contributory factor to this development. As a mobile technology, it has enabled grassroots efforts to gain a foothold in politics and democracy, from the most discussed Arab Spring to the Egyptian uprisings. Military strikes, coups, and much more can be read about on a Twitter feed. Thus, the influence of change through these mobile technologies was felt faster than through the traditional channels provided by regular news networks. Anyone could post information, in real time, so long as the bandwidth and technology networks were functioning in these countries.

More recently, we have seen the development of news sites or contributors that are trying to fit the mold of a society that is fast-paced, but which breaks the mold of the 24-hour news format or even the 30-minute news broadcast. TheSkimm, an e-mail news subscription service, which also has an app available for receiving their conversational and quirky style of news reporting, has 1.5 million subscribers to its daily newsletter, not including weekends

(Ellis, 2015). As Danielle Weisberg and Carly Zakin (2016), cofounders of theSkimm, stated,

> News is not only our career, it's our passion. Because of this, we have always been the go-to source for friends seeking the scoop on current events or breaking news. We soon realized three things: Reading the news is time consuming; Wanting to read the news is a hobby; lastly, not everyone has the time or interest.

They focused on a particular demographic to engage their audiences, in this case women who wanted a brief dose of news quickly. While different from using a regular social network, they used e-mail as their tool for communication, which drew criticism but proved to be the venue for accruing a mass audience.

These changes in the circulation of news and in civic engagement have brought on the idea that "in some ways, social networks have become the public ombudsmen for the news media by posting corrections, providing news, and reminding news agencies that there is a public to whom they should adhere and be accountable" (De Abreu, 2016). How does that translate to students and how can we bridge what that would mean in education circles or in the teaching of civic engagement to students?

Paul Mihailidis (2015), from Emerson College's Engagement Lab, advocates for the following four positions in developing participatory civic engagement:

- **Start Taking Sides**—In an increasingly ubiquitous media culture, media literacy can no longer be about critical thinking as an end goal. Critical thinking and critical expression, while central to the media literacy field, are not strong enough convictions to warrant the attention that media literacy needs in educational and policy arenas. Media literacy educators must start taking sides: directing critical thinking and expression towards goals, outcomes, and positions. Media literacy for health is about fighting against epidemics in obesity, heart disease, and other ailments that have plagued our time. The end goal must be to help reform issues in some way, shape or form. Otherwise, media literacy will continue to teach about issues, without teaching for issues. Media literacy analysis of political rhetoric about the environment, for example, should be directed at understanding sustainability, climate change, and the media's portrayal of these issues. While youth will come to their own conclusions, media literacy needs to position itself as the place where these decisions can be made, with confidence, humility, and tolerance.

- **Build Research to Support the Field**—To be able to take a more activist approach to our work—namely that media literacy is an essential pedagogy for leadership and engagement today—we need more research to back our work up. Media literacy scholars have been very proficient in writing about approaches to teaching and learning with media, but we have kicked our tires a bit to coalesce as a research-driven field in support of media literacy education. The academic journals now growing offer a start for the field, but too often our research gets subsumed by larger disciplines. Recent studies provide research that can be used to show the value, or need, for media literacy. This needs to be a bigger part of our mission and culture.
- **Connect Work to Policy**—This challenge may be the most difficult to engage with directly, but it's already started. Media education organizations and activists are proposing legislation to bring media literacy into schools in the United States, the E.U. is working to produce research and curriculum for its members, and in the Middle East, media education is beginning to sprout up in Beirut, Amman, and beyond. Media literacy advocates need to begin seeing their work as applied to help reform in formal and informal policy spaces.
- **Make Our Mission [Even More] Known**—Lastly, media literacy education needs to use its rich history to find its future voice. Media literacy education is well beyond needing to be justified anymore. As the term becomes more familiar, the media literacy community needs to have the scholarship, inquiry, and message ready. I'm less convinced this can happen if the community continues to rehash old divides, argue over best approaches, and think on a small scale (p. 56–57).

Civic engagement in education is hard to come by unless it is started with individual teachers or where teachers give voice to an issue that needs to be changed, especially in a time where education is being regimented. However, that has not changed the fact that teachers are the ones who are working with students and can see from the frontlines where work and development need to happen. This point has made engagement look different for teachers, yet it still happens in a way that impacts teaching and learning with the ultimate goal being impacting students in and out of the classroom. Here are some examples of where civic engagement can be seen in educational circles.

Edcamps

The unconference/Edcamp model, which originated in Philadelphia, came about when teachers nationally felt that professional development activities were not directed to helping them in growing their students' learning in the classroom. Most professional development work was directed to administrative work and data, none of which was helpful when teachers needed direct interventions to assist them with their work. Nationally, there have been over 1000 edcamps which have focused on a variety of different topics. There is now an Edcamp Foundation, (http://www.edcamp.org/), with a specific mission:

> Our vision is to create an international network of educators dedicated to participant-driven professional learning for themselves and others that accelerates student growth. Our vision is simple: innovative professional learning for educators. Our collective experience spans international education, professional development, technology, innovation, non-profit and for-profit business, and paramount to the work we do is input from others. Our charge is to help educators discover new ways of learning and teaching, reigniting their passion in the classroom and passing it on to students across the globe… .The conferences are participant-driven with the sessions developed and facilitated by the teachers who are attending and leverage social media to involve remote attendees. By adopting the un-conference protocol inspired by the Barcamp model, Edcamps provide access to high quality, personalized professional learning for all educators. Embracing the belief that fellow educators can be some of the most valuable professional development resources for their colleagues, the Edcamp movement has been recognized by countless organizations, including the United States Department of Education…the International Society for Technology in Education (ISTE)…and the Bill & Melinda Gates Foundation. (Edcamp Foundation, 2016)

Media Literacy Now

The policy leg for media literacy has never been more important and it has taken root with the organization Media Literacy Now. Their mission is "To spark policy change in every state and at the national level to ensure all K–12 students receive comprehensive media literacy education and skills" ("About," 2015). Since their inception, they have been able to have legislation passed or considered regarding media literacy and/or digital citizenship in the following states: Washington, Utah, New Jersey, Florida, Ohio, Illinois, New Mexico, Minnesota, California, Rhode Island, New York, New Jersey, Massachusetts, Connecticut, and Michigan. Their momentum is gaining as more concerned citizens from parents to educators to policymakers are wanting to see something in place that empowers teachers in a classroom to create and teach media literacy curriculum.

Within schools themselves, there is movement to have student voices heard so that they have an investment in the curriculum. This has been a slow process, and has not been met with open reception in administrative circles.

The role of technology in schools has more to do with the growing need for making students literate in a way that affords them an opportunity to communicate now and in the future, but even that has taken on a contested role when we look at the resources available to schools that fall under the scope of policy direction or misdirection, leading to the question of the digital divide.

Equal Access in Practice

> "Many people see technology as the problem behind the so-called digital divide. Others see it as the solution. Technology is neither. It must operate in conjunction with business, economic, political and social system" Carly Fiorina (2001), former chief executive officer of Hewlett-Packard.

The speed of Internet access has changed significantly in the last 20 years. The change has allowed people all over the world to be involved as participants in the creation of Internet websites as well as the production of online resources. Resources that were once in print are now available online. When we talk about books, we are also looking at e-books and online databases that support the proliferation of materials to the general public. While this change has been progressively happening in businesses, homes, and the general community, it has not been equal in schools or in the global community. Nor has the capability of access been equitable, which is why this chapter is significant in looking at how mobile learning is occurring within the United States and around the world. There are two main areas of concern: a digital divide in access and a digital divide in the progression of learning with and through the tools.

Divide in Access

For the longest time, when we looked at digital equity, we were looking at what was separating the world from being connected. Who had resources in the United States or in other parts of the world? Access points? Did they exist?

For the most part, throughout the United States there were always places that provided access points, from coffee shops to libraries. Many homes also carried their own infrastructure for having Wi-Fi access throughout. Schools, on the other hand, were problematic. Connection issues existed quite regularly. The reliability factor was always contentious. Many schools in the early 1990s had limited dial-up and that continued into the early 2000s. Services in homes were beginning to shift from dial-up to DSL service, but it took the last 5 years for things to really shift and change so that broadband was opened up. Yet schools were still being held back. Money for developing older buildings that didn't have the wiring capable for moving technology forward became a part of the regular conversation. In poorer urban schools, access was almost a moot point, with little funding in place and even less for technology-related items that were not considered a necessity. Teachers in various communities were asked to respond to whether the digital divide existed in their districts. The digital divide is interpreted differently from place to place, which is evident in the following statements.

> The digital divide is alive and running. I teach currently in an urban setting. My school is PreK–8 and just like most of the teachers in my building I have noticed that almost every student from grades 5–8 and especially the seventh and eighth have smartphones. They pretty much use them for social media, music, and gaming. Even though there is a district policy prohibiting the use of cell phones in the school during school hours. Yet, the students don't know or even care that they can use their phones as a small computer if they are taught the right way of its potential. Shoot, I am still learning my Galaxy S5 and its full capability. I am still a BlackBerry fan. Anyway, there is a disconnect with our society and there is a digital divide. Now I might be a physical education teacher but when trying for my students to do some "gym homework," I asked a few classes to use YouTube so they could see videos on how to do perfect pushups, 90-degree arm-bend pushups. After making this announcement my co-teacher informed me that there was a great possibility that many of the students did not have Internet access. I was embarrassed and shocked at the same time, I did not even think of the digital divide. But it is alive and well. The next class [that] I asked at the end of the period [to do] the same assignment, this time I started by saying, "who has Internet access at home?" I was again shocked by the almost 55%–60% of the class [who] DID NOT have access. (Classroom Teacher A)

> Most of the schools in which I had taught have been in communities where the majority of the families have internet access within their homes and have at least one functioning computer or device, in addition to any cell phones or smartphones. I have also been in schools in the Bronx and in Bridgeport where the only Internet access the students had was at school or at the public library, and many of the students did not have computers at home, although many of them did have cell phones

and iPods. In looking at the digital divide and in thinking about the readings, presentations, and videos we had this week, there appear to be several different factors in the divide. The divide is not simply having a device or not having a device. It also pertains to having access to broadband Internet and knowing how to use the existing resources. (Classroom Teacher B)

In my community of practice, digital equity falls on the side of the haves, who choose to have (or not) at the school level and in the homes [in which] they live. The money and resources are there, so it just becomes a question of what ICTs (information and communications technology) do we want and when and how do we want them. I work at the elementary level and it has been frustrating because the money is there but the town chooses to allocate most of the technology funding to the middle and high school. The town is extremely old fashioned and doesn't understand why elementary students need more technology. Although, I would love more technology at my school, I know that we are blessed. The previous district I worked in was in a city, and while some of the magnet schools were overflowing with resources and technology, most of the schools were severely lacking in ICTs. At home, many students had access to no technology at all. In this district, one school I worked in was a PreK–8 school. I primarily worked with the students in grades 6–8. One thing I noticed is that the kids carried their cell phones everywhere and were constantly using these devices. Many had no backpacks, no pencils, no books, but they had their cell phones. I'm not sure if this had more to do with age than income level. I currently work in a middle/upper class elementary school and I've never had the opportunity to work in a middle school in this type of demographic, but I'm guessing that the difference is that the middle/upper class kids may use their cell phones all the time but they also have a NOOK, an iPad, or desktop computer available at home to learn and do work. Most of the inner-city kids did not have these ICTs available at home to learn and do work and the cell phone was their only means of online connectivity. (Classroom Teacher C)

In my community of practice, the majority of my elementary school students have access to the Internet on at least one device at home. Most students are able to utilize a desktop or laptop. However, the difference between students with and without computer and Internet access has grown more noticeable as learning has become more Web-based in even just the past few years. My district has moved over to Google Apps for education, and all third-through-twelfth-grade students have educational access to Gmail and Google Drive. As such, teachers now often require students to do their homework on Google Docs and "share it" with them electronically. Students without devices that have easy-to-use keyboards and word processing functions are at a clear disadvantage. Similarly, the district has Web-based math and literacy accounts for students. While some of these Web 2.0 programs can be accessed on mobile devices (iPhones and iPads, tablets), at times their functionality is limited on these platforms. Moreover, as a computer teacher, we instruct our students in keyboarding, proper typing, finger placement, etc. Students have accounts

> on Web-based typing programs. When teachers assign "typing" practice for homework, students without Internet access or without computers with keyboards are at a significant disadvantage. Therefore, I do think that access in and of itself continues to be an issue to varying degrees in all communities. One way our district is seeking to address it is by looking to provide students at the high school level with a Chromebook if they cannot afford a computer of their own. In the coming years, high school students will be expected to bring a laptop computer (Mac, PC, or Chromebook) to and from school daily. This will ensure that students do not just receive a device, but they learn how to use it. Thus, access will be coupled with education. (On a separate note, I am not sure how the district plans to address the issue of families who lack complete Internet access at home.) (Classroom Teacher D)

These teachers represent an audience from both suburban and urban communities in the United States. On a worldwide scale, access has been a point of contention, and certainly the divide is evident in many places around the world, especially in Africa and some parts of Asia. By the same token, access has moved faster in other parts of the world such as South Korea. However, in terms of learning possibilities, whether it is the United States or abroad, educators are facing the same uncertainties.

Divide in Learning

Jenkins (Jenkins et al., 2006) would call this the "participation gap—The unequal access to the opportunities, experiences, skills, and knowledge that will prepare youths for full participation in the world of tomorrow." The upside and downside of technology sometimes has less to do with the actual divide of equipment or resources and more to do with knowledge once you have the tools in hand. As was recently stated in an education article,

> The phone could be a great equalizer, in terms of giving children from all sorts of socio-economic backgrounds the same device, with the same advantages. But using phones for learning requires students to synthesize information and stay focused on a lesson or a discussion. For students with low literacy skills and the frequent urge to multitask on social media or entertainment (which often seem to go hand in hand), incorporating purposeful smartphone use into classroom activity can be especially challenging. The potential advantage of the tool often goes to waste. (Barnwell, 2016)

Many educators are not equipped with the knowledge of using mobile technologies for learning or instructional purposes. The entertainment value of the tool has displaced the understanding of the value of the tool or application when investing in learning gains. Not because there is not a need or a desire, but because there has been little professional development time afforded to teachers in order for them to be successful in the classroom with the tool and the learners they face. This is reflective of the comment made by George Lucas (in Daly, 2004),

> When people talk to me about the digital divide, I think of it not being so much about who has access to what technology as who knows how to create and express themselves in this new language of the screen. If students aren't taught the language of sound and images, shouldn't they be considered as illiterate as if they left college without being able to read or write?

Learning with digital technologies is a literacy learning, and it can be as basic as understanding the way in which text flows on a screen or as complex as understanding the most important piece that is critical thinking. Critical thinking necessitates an understanding of how the educational material or tools are put together, how information is transmitted, and how it can be perceived by the user and the receiver.

The article "Bridging the New Digital Divide" (Day, 2013) makes the key point that the mobile devices, or any other technological devices, for that matter, are worthless without a competent adult supporting their use. The National Education Technology Plan (2016) also makes a point that there are two variations of the digital divide, and the more current one has to do with learning: "We have to be cognizant of a new digital divide—the disparity between students who use technology to create, design, build, explore, and collaborate and those who simply use technology to consume media passively" (U.S. Department of Education, 2016, p. 18).

The other issue faced with the idea of the digital divide is the digital wasteland—time that is lost because students are so invested in their technologies for more passive elements than the ones that are engaging, participatory, civic, or instructional. The hours lost to items on digital technologies that have little impact are not lost on educators and are certainly a concern when it comes to mobile technologies and digital learning. It is a point that we can all be accused of within any given day, but it is more particularly concerning when it comes to young adults, when what we seek is more attentive learners in the classroom and in our society.

While providing accessibility to all learners is a provision of having mobile technologies and is supported by various learning principles, a path to that pursuit must be in place. As Yildiz (2016) stated, we must:

- Provide multiple means of representation so that students can approach information in more than one way.
- Provide multiple means of expression so that all students can demonstrate and express what they know.
- Provide multiple means of engagement to stimulate interest in and motivation for learning.

The final goal is to empower learners to be critical thinkers and problem-solvers despite the variations of digital divides—in effect, to be consumers and producers of information using mobile technology tools, which enables true engagement in learning and eventually within civic life.

References

About. (2015). Retrieved from http://medialiteracynow.org/about-us/

Alexis de Tocqueville Quotes. (1835). *Quotes.net*. Retrieved from http://www.quotes.net/quote/60057.

Barnwell, P. (2016, April 27). Will giving greater student access to smartphones improve learning? *The Hechinger Report*. Retrieved from http://hechingerreport.org/will-giving-greater-student-access-smartphones-improve-learning/

Benkler, Y. (2006). *The wealth of networks: How social production transforms markets and freedom*. New Haven, CT: Yale University Press.

Critical thinking explained to students. (2016). Retrieved from http://www.educatorstechnology.com/2016/05/critical-thinking-explained-for-students.html

Day, L. (2013, January 3). Bridging the new digital divide. *Edutopia*. Retrieved from http://www.edutopia.org/blog/bridging-the-new-digital-divide-lori-day

Daly, J. (2004, September 14). Life on the screen: Visual literacy in education. *Edutopia*. Retrieved from http://www.edutopia.org/life-screenDe Abreu, B. (2016). Global media events and moments. In B. De Abreu & M. Yildiz (Eds.), *Global media literacy in a digital age* (pp. 39–49). New York, NY: Peter Lang.

Dumlao, M. (2005). *The West Wing* legacy: Rescuing democracy and civic engagement. *Critical Approaches to TV*. Retrieved from http://www.michaeldumlao.com/PDFs/West%20Wing%20Legacy.pdf

Edcamp Foundation. (2016). What is edcamp? Retrieved from http://www.edcamp.org.

Ellis, J. (2015, August 18). How The Skimm's passionate readership helped its newsletter grow to 1.5 million subscribers. Retrieved from http://www.niemanlab.org/2015/08/how-the-skimms-passionate-readership-helped-its-newsletter-grow-to-1-5-million-subscribers/

Fiorina, C. (2001, April 25). Speeches: Crossing technological divides. *Confederation of Indian Industry Conference*. New Delhi, India. Hewlett-Packard Development Company. Retrieved from http://www.hp.com/hpinfo/execteam/speeches/fiorina/ciic_01.html.

Jenkins, H., Purushotma, R., Clinton, K., Weigel, M., & Robinson, A. (2006). *Confronting the challenges of participatory culture: Media education for the 21st century*. Chicago, IL: The MacArthur Foundation. Retrieved from https://mitpress.mit.edu/sites/default/files/titles/free_download/9780262513623_Confronting_the_Challenges.pdf

Martin, J. N., & Nakayama, T. K. (2010). *Intercultural communication in contexts*. New York, NY: McGraw-Hill. Retrieved from http://www.rasaneh.org/Images/News/AtachFile/15–8-1390/FILE634561743619907963.pdf.

Mihailidis, P. (2015). Media literacy education: The new civic currency. *The Journal of Media Education*, 62(1&2), 56–59.

Perrin, A. (2015, October 8). Social media usage: 2005–2015. *Pew Research Center*. Retrieved from http://www.pewinternet.org/2015/10/08/social-networking-usage-2005-2015/.

Purcell, K., Heaps, A., Buchanan, J., & Friedrich, L. (2013). *How teachers are using technology at home and in their classrooms*. Washington, DC: Pew Research Center's Internet & American Life Project.

Richtel, M. (2012, May 29). Wasting time is the new digital divide. *The New York Times*. Retrieved from http://www.nytimes.com/2012/05/30/us/new-digital-divide-seen-in-wasting-time-online.html?_r=0

Tocqueville, Alexis de (1835). *Democracy in America*, Volume I. Saunders and Otley.

U.S. Department of Education. (2016). *Future ready learning: Reimagining the role of technology in education*. Retrieved from http://tech.ed.gov/files/2015/12/NETP16.pdf

Weisberg, D., & Zakin, C. (2016). Who are theSkimms? Retrieved from http://www.theskimm.com/about

Yildiz, M. (2016). Practice to action: World savvy teachers. In B. De Abreu & M. Yildiz (Eds.), *Global media literacy in a digital age* (pp. 219–237). New York, NY: Peter Lang.

Zacchetti, M., & Vardakas, P. (2008). A European approach to media literacy. In U. Carlsson, S. Tayie, G. Jacquinot-Delaunay, & J. Tornero (Eds.), *Empowerment through media education* (p. 119). Göteborg, Sweden: The International Clearinghouse on Children, Youth and Media.

· 4 ·

PRIVACY, STUDENT DATA—KNOWLEDGE AS EMPOWERMENT

> A child born today will grow up with no conception of privacy at all. They'll never know what it means to have a private moment to themselves, an unrecorded, unanalyzed thought. And that's a problem because privacy matters; privacy is what allows us to determine who we are and who we want to be.
>
> —SNOWDEN, 2013

> Privacy is an inherent human right, and a requirement for maintaining the human condition with dignity and respect.
>
> —SCHNEIER, 2015

The greatest discussion piece in recent years surrounds what or who are we exposing students to when offering new technologies in the classroom? For the most part, there has been an assumed belief that students' data was protected, and that students were protected from outside Internet companies. This belief has been shared by administrators in schools, parents, and even the students. It was, in part, manufactured by a perceived understanding based on various United States laws and, in general, the idea that students were off-limits to the forces of some of the advertising as well as other methods of lurking that most adults expect when online. Unfortunately, this assumption proved to be a mistake on all counts—from Google, the most well-known of the offenders, who had been scanning and data-mining student's e-mails from their "free" school Gmail accounts, to other educational tool companies, including testing providers who were found to be breaching a United States law called the Family Educational Rights and Privacy Act (Hern, 2014; Herold, 2014).

What is Private?

This question is difficult to answer with the amount of technology that is open and available to people around the world. The subject of privacy has taken to the airwaves, news magazines, and more. The concern is more than just about the information that adults put out on the web, but also the information our students and children leave behind. The thoughtful engagement of the process of information is missed by many, as is the vital competency of media literacy.

As adults we tend to have a better awareness of what "mediated messages" means. We question more of what we see and hear. We tend to understand how data is collected, manipulated, and how it affects us if our information is freely available to the public. Students, especially our youngest ones, and middle-school age most particularly, tend not to have this knowledge base. Without the necessary learning about a digital footprint and digital citizenry, they forget that they are not alone in the world. Their online lives can be seen by many; can be absconded, stolen, or repurposed in a way that is dangerous; and can lead to lifelong consequences. Besides students freely giving information, the other and greater issue is the information that is taken by companies for data collection purposes. The information taken in this way is more insidious, less obvious, and much more purposeful, whether it is by companies such as businesses who follow students, or educational groups whose purpose is to gather data for the analysis of student knowledge.

Is Privacy a Luxury Good?

The real concern is this: how much are all of us willing to pay for the option to have some privacy when we live in a world that has been permeated by technology tools that we want to have and use so much? Our students face the same dilemma, but in many cases are not asked or given a choice, and most are unaware that they should be concerned. For quite some time, technology distributors counted on that fact until the Student Privacy Pledge was initiated and parents began raising their concerns publicly about how, as well as where, their children's information was being kept and distributed. The reality is that so much of all of our private information is already available and difficult to keep confidential. This question of privacy as a luxury good is a part of my current research as well as a part of the growing debate in schools.

Parents, Common Core, and Testing

As a child does not have world or life experience, the role of the parents is immeasurable, especially when it comes to protecting their rights. This can be most noted in what took place in the past year with the data collection and the data mining of students' academic records along with the "opt-out" movement on testing. Parents became aware of the systematic collection of information that was taking place from the amount of testing that was being done in the name of bettering education. The issue was not so much how the data was collected, but how third parties were given access to the data, and that was not a part of the original terms that were understood by parents. This, particularly, has reared its ugly head most recently in New York State with a company called inBloom. InBloom was a non-profit organization funded by the Gates Foundation and Carnegie Corporation whose purpose was to collect testing data (Kharif, 2014; Vijayan, 2014). As of April 2014, the company closed down because of parental protests and demand for transparency. This incident was the first example of a grassroots effort, led by parents, to combat the way in which student data and privacy is handled at the academic level.

Parents are a force to be reckoned with when they balk against what they consider to be some of the most grievous ways in which data is being taken from their children without the parents' consent, and certainly with even less knowledge from the children. The "opt-out" movement has followed along the same lines as parents became more and more fed-up with how testing was dominating education to the point of hurting their children's ability to succeed in school. Many parents considered the testing movement, along with the way in which the Common Core State Standards (CCSS) were implemented, another way for gathering information at the expense of their children's education. Over the past year, several states have decided to drop out of the Common Core and reconsider the standards by which their students will be evaluated. Along with New York State, the opt-out movement impacted Colorado, New Jersey, and California (Strauss, 2015). High school students even walked out of high schools, refusing to take the test—a test which, in fact, had little meaning to them as they moved toward college or career (Evans & Saultz, 2015). These incidents have created further testing modifications and changes. In Connecticut, high school students will no longer be taking the SBAC, another standardized test, and instead, students in Grade 11 take the SAT (Harris, 2015).

The inBloom incident, with parents not wanting their children tested as part of the new Common Core State Standards, has become quite controversial. Some would say that movements are impeding educational growth, which was, in fact, a comment made by former U.S. Secretary of Education Arne Duncan (Evans & Saultz, 2015). The reality is that parental efforts were a success and certainly should be highlighted for others in the worldwide community as a model for what can be done when concerns of data breaches are evident. Moreover, what is most apparent is that the idea of privacy appears to be misunderstood and undefined in schools. There is a need for data and for the use of technology in schools, as the National Association of State Boards of Education stated,

> When teachers and schools use data and technology to tailor instruction to individual needs, students benefit through enriched, accelerated learning. Teachers and schools can use education data to measure whether particular teaching methods are promoting student learning. State policy makers can use date to make judgments about the effectiveness of standards implementation and then improve policies or allocate additional state funds or technology support in response. (Vance, 2016)

Unfortunately, schools collected data before putting policy in place for protecting the data. Moreover, many administrators and educators are unaware or ill-informed about products used to collect data about their student body under the guise of assessment purposes. These are hard lessons learned by schools who have had data breaches or have left their data vulnerable to attack.

Questions Teachers Ask Themselves

Through graduate education courses and various presentations given this past summer, it became very obvious that there were a variety of concerns and questions from educators, parents, academics, and the general public. Teachers in particular asked:

- How can teachers maintain boundaries for students between their academic lives and their personal ones if students are using the devices, networks, and accounts that allow for data tracking?
- How do we educate students about their "privacy" (and lack thereof) and how to protect it?
- How do teachers maintain boundaries between their professional selves and their personal lives with regard to their use of technology?
- Who makes certain these records are disposed of in an efficient and effective manner?

As a teacher mentioned in conversation, "What is next to impossible to monitor, in my opinion, is the massive amount of student data that is now collected and stored. NCLB (No Child Left Behind) and SBAC (Smarter Balanced Assessment Consortium) call for the collection and collation of student data." Indeed. The inBloom incident is evidence of these types of concerns, but what about the idea of privacy in general? How do teachers understand it for themselves?

In a recent graduate class with current and future teacher educators, this very question was posed. Their comments demonstrated a widening concern for data mining within education:

> "I am concerned and surprised to find out that student data is not only unprotected, but also actively collected and sold" (Secondary Educator).

> The ever-increasing advancements in technology bring more and more attention to privacy. One can think that they know all the ins and outs of any given form of technology, but at the end of the day there is still going to be someone out there who knows more than you. It's very important to educate yourself on all the different ways you can prevent predators from making you a victim. There is a wide variety of ways that a lack of privacy can harm you. These include getting your identity stolen, all the way to someone falsifying information about you on the web. (Graduate Student)

The other issue that many educators think about is that they know very little information about laws related to privacy in and outside of schools, especially as technology changes are occurring so quickly. For example, many teachers have heard the terms FERPA or COPPA, but don't know what they mean. Just for clarification, FERPA stands for Family Educational Rights and Privacy Act. The goal of FERPA is to protect the confidentiality of education records. Those records include "any files, documents, or other materials that are maintained by an educational agency or institution or by a person acting for such agency or institution and contain information directly related to a student" (Ritvo, Bavitz, Gupta, & Oberman, 2013). COPPA stands for Children's Online Privacy Protection Act. COPPA delineates "privacy standards and obligations for online service providers that either target children or knowingly collect personal information from children under the age of 13" (Ritvo et al., 2013).

The question then becomes who will oversee the records and make sure that schools are in compliance and that companies associated with schools are also in compliance? Teachers are overloaded with content curricula and new requirements each year and these concerns weigh heavily as new methods of data collection are introduced.

Challenging Privacy Concerns

In between all of the concerned educators and parents are also those who question whether privacy is even an issue or who do not see the way in which data is managed as problematic. In a graduate class, future educators were asked to review and comment on the Glenn Greenwald (2014) TED Talk that discusses the idea of privacy in detail. Glenn Greenwald is best known as a Pulitzer-Prize-winning journalist and one of the contributors to/collaborators in *Citizenfour*. In the short, 20-minute talk, Greenwald discussed the issues he has with what is happening when our privacy is invaded.

The talk entitled "Why Privacy Matters" goes on to detail the concerns with the misappropriation of data. In one section, Greenwald (2014) states that he challenges all the audiences he has spoken to that they really consider this idea of privacy. When he encounters someone who states that they are not worried about the invasion of their privacy, he leaves them his e-mail address and asks them to send him their passwords to all their e-mail accounts. As he states, "Not a single person has taken me up on that offer. I check."

One student responded to the TED Talk by stating:

> I do, in fact, have concerns about privacy and the implications with technological "progress." My concerns exist on several levels. I was familiar with Glenn Greenwald (having recently watched the *Citizenfour* documentary) and have shared his concerns about government surveillance of the public at large. The idea of living our lives in complete transparency seems idyllic, until we realize what that really means. In addition to needing privacy for a sense of creative liberated freedom, we have to consider that it's not just our information that is monitored, but rather bits and pieces that

> are perceived and analyzed leading to conclusions that may or may not be correct! It's scary to think of how things about us out of context could change their reality.

The reality is that big data violations and privacy breaches are felt by all. Responses to the challenge of privacy concerns lead to an awareness that media literacy education is needed for both educators and the students they teach.

Media Literacy Takes Action

The role of media literacy education in relation to privacy and data is incredibly important. The framework provided by the Center for Media Literacy provides educators and students with a way of understanding how systematic companies are with obtaining information and how we, as consumers, need to be responsible participants in this ongoing dynamic of change. The case of privacy reinforces why media literacy education is essential and must be taught in schools as technology companies continue to introduce ways in which they can invest in schools and garner education participation. At the same time, the idea of "free" in education must be reevaluated and critically considered when it comes to schools and the use of technology by students, teachers, and administrators.

There is no question that the issue of privacy has been shaped globally. Our European counterparts have been fighting to get their information back, especially from companies such as Google, with success. France, Germany, and the UK have all had lawsuits against Google stating that the company has violated data protection laws (Essers, 2013; Sterling, 2013).

The growing concerns have related back to the laws whereby citizens wanted information about themselves to be disposed of online. Citizens in the EU began the discussion of concerns related to the struggle to remove items that had been posted online. Appealing to Google had no end result. In March of 2010 a Spanish citizen filed a complaint with the Spanish government after a notice about his repossessed home came up on Google (January, 2016). He had gone to Google and asked for its removal because he felt that this information was private, and, in effect, that he had suffered both personally and financially because that information had been made public (Powles, 2014). The Spanish court placed the case in front of the Court of Justice of the European Union asking, "whether the E.U.'s 1995 Data Protection Directive applied to search engines such as Google" (European Commission, 2014).

The discussion that ensued became known as the "Right to be Forgotten," which was later referenced as the "Right to Erasure." The challenge began to circle around where do you draw the line as to what is permitted online, and who decides where or if it should be removed? The end result of the case was a new law described as Article 17, stating the following:

Article 17 Right to be Forgotten and to Erasure

1. The data subject shall have the right to obtain from the controller the erasure of personal data relating to them and the abstention from further dissemination of such data, especially in relation to personal data which are made available by the data subject while he or she was a child, where one of the following grounds applies:
   (a) the data are no longer necessary in relation to the purposes for which they were collected or otherwise processed;
   (b) the data subject withdraws consent on which the processing is based according to point (a) of Article 6(1), or when the storage period consented to has expired, and where there is no other legal ground for the processing of the data;
   (c) the data subject objects to the processing of personal data pursuant to Article 19;
   (d) the processing of the data does not comply with this Regulation for other reasons.
2. Where the controller referred to in paragraph 1 has made the personal data public, it shall take all reasonable steps, including technical measures, in relation to data for the publication of which the controller is responsible, to inform third parties which are processing such data, that a data subject requests them to erase any links to, or copy or replication of that personal data. Where the controller has authorised a third party publication of personal data, the controller shall be considered responsible for that publication.
3. The controller shall carry out the erasure without delay, except to the extent that the retention of the personal data is necessary.
   a. for exercising the right of the freedom of expression in accordance with Article 80;
   b. for reasons of public interest in the area of public health in accordance with Article 81;

c. for historical, statistical and scientific research purposes in accordance with Article 83. (European Commission, 2014)

The dialogue and the court case resulted in several articles being written worldwide on the subject. Some people were very much for this right and others felt that it would eliminate a disclosure of history that was vital to how events had transpired. Some items appeared benign and not so essential to have removed, but others, especially in the case of personal attacks or even in the case of bullying, reared their heads from the perspective of the victim versus the victimizer. These concerns continue to be voiced despite the legislation in effect.

Legislative direction is just beginning to take shape in the United States, but the process is slow. California is one of the first states to have passed a law protecting digital privacy rights. The Electronic Communications Privacy Act

> bars *any* state law enforcement agency or other investigative entity from compelling a business to turn over any metadata or digital communications—including emails, texts, documents stored in the Cloud—without a warrant. It also requires a warrant to track the location of electronic devices like mobile phones, or to search them. (Zetter, 2015).

Within education, this is still where the greatest work needs to be done so that we create discerning learners and future citizens who are responsible data managers and producers of the future.

Student Privacy Pledge

The Pledge is a list of 12 commitments that school service providers have made to assert and acknowledge that K–12 student data is maintained in a secure, private, and responsible framework. President Obama has endorsed the Student Privacy Pledge, citing it "as a strong means of protection for student personal information" (The White House, 2015). The Student Privacy Pledge was developed in conjunction with the following parties: the Future of Privacy Forum (FPF) and the Software & Information Industry Association (SIIA). As it currently exists the Pledge states:

We Commit To:

- ✘ Not collect, maintain, use, or share student personal information beyond that needed for authorized educational/school purposes, or as authorized by the parent/student.
- ✘ Not sell student personal information.
- ✘ Not use or disclose student information collected through an educational/school service (whether personal information or otherwise) for behavioral targeting of advertisements to students.
- ✘ Not build a personal profile of a student other than for supporting authorized educational/school purposes or as authorized by the parent/student.
- ✘ Not make material changes to school service provider consumer privacy policies without first providing prominent notice to the account holder(s) (i.e., the educational institution/agency, or the parent/student when the information is collected directly from the student with student/parent consent) and allowing them choices before data is used in any manner inconsistent with terms they were initially provided; and not make material changes to other policies or practices governing the use of student personal information that are inconsistent with contractual requirements.
- ✘ Not knowingly retain student personal information beyond the time period required to support the authorized educational/school purposes, or as authorized by the parent/student.
- ✔ Collect, use, share, and retain student personal information only for purposes for which we were authorized by the educational institution/agency, teacher, or the parent/student.
- ✔ Disclose clearly in contracts or privacy policies, including in a manner easy for parents to understand, what types of student personal information we collect, if any, and the purposes for which the information we maintain is used or shared with third parties.
- ✔ Support access to and correction of student personally identifiable information by the student or their authorized parent, either by assisting the educational institution in meeting its requirements or directly when the information is collected directly from the student with student/parent consent.
- ✔ Maintain a comprehensive security program that is reasonably designed to protect the security, privacy, confidentiality, and integrity

of student personal information against risks – such as unauthorized access or use, or unintended or inappropriate disclosure – through the use of administrative, technological, and physical safeguards appropriate to the sensitivity of the information.

- ✔ Require that our vendors with whom student personal information is shared in order to deliver the educational service, if any, are obligated to implement these same commitments for the given student personal information.
- ✔ Allow a successor entity to maintain the student personal information, in the case of our merger or acquisition by another entity, provided the successor entity is subject to these same commitments for the previously collected student personal information. (Student Privacy Pledge, 2015)

There are over 200 companies who have signed the Student Privacy Pledge, from Apple to Blackboard to Turnitin. All of the companies on the list are considered school service providers, which means they are interacting in some way with schools or students in the capacity of their education:

> "School service provider" refers to any entity that: (1) is providing, and is operating in its capacity as a provider of, an online or mobile application, online service, or website that is both designed and marketed for use in United States elementary and secondary educational institutions/agencies and is used at the direction of their teachers or other employees; and (2) collects, maintains, or uses student personal information in digital/electronic format. (Student Privacy Pledge, 2015)

The full list of signers can be found in the Appendix of this book.

Privacy and Data in Transition

Like any big issue, privacy and data are in a place of transition and conflict, one that will have repercussions in the future, whether for schools, the federal government, Internet companies, or other mobile application developers who handle encrypted communication. The most notable case of late deals with Apple and the FBI who had a major dispute over the ability to get data from an iPhone that was used in the San Bernardino attacks. The FBI wanted Apple to help unlock the phone and work around the encryption. Apple refused. It has said that encryption is key to protecting user data from hackers (Kharpal, 2016). Apple is looking at it from a larger perspective of its users, in terms of how it can continue to provide services that will be secure, especially as it is one of the reasons why people use this product. For this purpose, the FBI was looking at a criminal case, but the issue goes beyond what is being said in the moment. The issue of power and the dialogue of where the court of public opinion falls on data privacy is placing people in a juxtaposition where the person or business of who is right or wrong is not so clear.

John Oliver (2016) from HBO's *Last Week Tonight* explains in a satirical yet amusing way both sides of the problem. However, at the end, his interpretation provides listeners with an opportunity to see the larger context of where this could be going:

> Whatever happens in this case will have ramifications because the FBI ultimately wants Apple and the entire tech industry to have its encryption always be weak enough that the company can access customers' data if law enforcement needs it so it might be the iPhone today, an Android phone tomorrow, and a BlackBerry the day after that, assuming that the day after that is in 1998.

Since his show aired, the FBI was able to break the encryption internally. In turn, they dropped the case against Apple, but the conversation isn't over. The FBI will not release how it was able to work the system, and Apple most assuredly is working to close the backdoor loophole that was discovered. Perhaps the fact that it was taken out of the court system in this way was a boon because the reality of what could have happened if the case had gone against Apple may have posed potential international issues, where countries such as Russia and China would also want similar access if the U.S. government had won the case (Smith, 2016).

Digital Citizenship

As part of the discussion of privacy, an opportune moment is made available to provide students with tools to help them govern their online presence and help them in cultivating their digital footprint. In the United States and around the world this topic is one that has been discussed and is beginning to be introduced as a curriculum in schools.

At SXSW 2016, the topic of digital citizenship was the focal point of several presentations, including a core conversation that was part of a collaborative project presented between *Cable Impacts Foundation* and myself. The question that framed the purpose of the session: Are adults making it better or worse? The forum was designed to build a holistic view and understanding of national and international conversations/concerns bridging policy to practitioners on the issues of digital citizenship. At the same time, the gathering raised awareness in order to bring greater understanding to a global education topic. The group in attendance was asked to come to a consensus about the most pressing concerns related to students and their digital lives, acknowledging questions/concerns/reflections, with the hope of extending the conversation, providing knowledge, and offering the opportunity to critically engage in the current dialogue. Some of the intended outcomes were to engage/spark a conversation with other folks in order to create a living/working document enabling participants to walk away with a tangible resource that they can take back to students, parents, educators, or other interested parties. Questions that were explored included:

- Problems with the message delivery/adults who are providing the lessons?
- Do as I say, but not as I do.
- Consider what world do most adults live in versus the ones of students?
- How did that shape the message?
- Did it, in fact, skew/distort the digital citizenship message?
- Are educators/parents/adults in general a part of the problem?

The session began with an introduction to a variety of current news articles that focus on the concerns surrounding students' behaviors online. The group of 100 participants was divided and the discussion ensued. The use of Google Docs, the most accessible technology tool for this purpose, accessed on a variety of mobile devices, allowed participants to comment and respond while having interpersonal conversations. Their responses to the questions posed are recorded here:

- Are we imposing our views/belief systems into a world in which we may not fully participate?
    - o Yes! Yes!
    - o Is this bigger than just "technology"? If we expect kids/adults to be good Digital Citizens, shouldn't that assume that we also expect them to be good "real life" citizens? Maybe there are cultural problems that are allowing for misbehavior/bad decision-making,

which are stimulating a need for something specific that we call "Digital Citizenship."

- Am I, as an adult, trying to control a student's digital footprint in a way that isn't functional to the student?
    - Students don't understand the long-term implications. How do you allow students to make mistakes that don't follow them for a lifetime?
    - Consider giving students an opportunity to create a positive digital footprint through development of a digital portfolio/résumé starting in elementary school.
    - Is it possible that in the near future, most people will have things online we are not proud of? Maybe standards will relax because so many will have an online "past." There's probably a fortune to be made if someone could invent a tool to wipe the Internet of all online references we don't like about ourselves.
- Are there problems with the message delivery from the adults who are providing the lessons? (Such as proper modeling; do as I say, but not as I do.)
    - Yes, relevance is an issue. Too many teachers/adults rely on curriculum that is out-of-date the moment it is published. Videos reference tools that kids don't use anymore or scenarios that don't reflect the truth about what kids are doing right *now*.
    - Sometimes presenting a "sanitized" web presence doesn't empower kids to recover from past "oopsies"—important to show good decision-making after an "oopsie" and help kids to cultivate a positive online profile.
    - It depends on the knowledge/comfort level of the adult.
- Consider what world do most adults live in: How does that shape the message?
    - Need to consider socio-economic issues/disparity when it comes to accessing tech and understanding the inherent issues.
    - Adult/parent education is just as important. Show them how to engage with their kids relating to online activity.
    - A lot of this starts with the school library.
    - The relationship between the student and adult is important to the impact the message has. Parent/child vs teacher/child may have different impact—other adults in a student's life may have

the ability to reach them better/more effectively (counselor, other family member, church leader, etc.).

- How can you make digital citizenship a habit, an ingrained mindset, action, and behavior?
  - o Policymakers build a climate that's supportive of the instruction of digital citizenship.
  - o Integrate into curriculum.
  - o Just as we teach playground safety and bus safety, we can also teach Internet/online safety and proper use related to digital citizenship.

In an oral discussion, participants stated they had not considered that their instruction was based on their own perceptions of how things should be handled in a digital age. One participant voiced their idea that digital citizenship should be taught, similar to sex education in schools. This brought about the discussion of the inoculation method of teaching, which has not been effective. In the meantime, the document concerning this topic continues to grow with further questions and ideas.

The reality is that the only way to keep students safe online is to keep them offline or off the grid. Unfortunately, in this day and age, that is not reasonable or even a sound method for handling the inevitability of students using the Internet, social network sites, or being faced with an online environment without having any prior knowledge of how to conduct themselves.

Several organizations have begun to share their curricula or work involving digital citizenship. Among them:

- Common Sense Media: https://www.commonsensemedia.org/kids-action/issues/privacy-and-safety/school-privacy-zone
- iKeepSafe: http://ikeepsafe.org/educators_old/schoolprivacy/

Another non-profit organization, CyberWise, published in the *Huffington Post* "10 Questions Parents Should Be Asking Regarding Internet Safety." These questions, while directed at parents, have purpose for educators and ask us to consider all aspects of the digital life from reputation to media literacy. The questions are worth discussing at all levels and with multiple stakeholders who have an investment in learning and in student education.

10 Questions Parents Should be Asking Regarding Internet Safety

- Reputation Management. Do your students know how to be proactive managers of their "digital footprints"? Do they (this includes adults and students) even know what a "digital footprint" is? Do they know it's how future friends, employers, and college recruiters will likely judge them? Do they understand that their "footprint" includes not only what they post about themselves, but also what others post about them? This includes when they are "tagged" by friends and/or followers. It also includes all those baby pictures mom began sharing years ago.
- Online Privacy. Personal information is the currency of our day, and understanding how, why, and when to give it away is important knowledge for students. So ask yourself how much your students know about the Privacy Policies they agree to before joining all those social networks they use.
- Online Relationships. Do the adults in your community set boundaries and expectations that help students experience positive online relationships? Are students being taught how to avoid and/or prevent cyberbullying, sexting, online predators or other negative and potentially dangerous online relationships? Do they know how to be "upstanders" if and when they (and they will) observe others being treated cruelly online? Does your school provide information and help to students who experience cyberbullying? Has your school advised students how to handle a request from another student for a "nudie" (a not uncommon occurrence by the way)? Do they know what to do if they receive a sext message?
- Online Security. Are your students savvy about identity theft, viruses, malware? Do they know how (and why) to create strong passwords? Have you taught them how to recognize and what to do in the event of a phishing attempt?
- Ethical Decision-Making. Since nearly every online action requires ethical thought (i.e., "Should I post an embarrassing photo of a friend? Or download a movie I haven't purchased"?), young people should be practiced in making ethical decisions online and off. Most children can't even engage in ethical thought until 12 years of age or later; do your parents know this and thus keep their children from using social media until then? Are your students taught how and why to avoid

possible illegal activities like plagiarism, copyright infringement, or hacking?

- Time Online. What about setting healthy limits and expectations for media use? Is your community familiar with the American Academy of Pediatrics' media use recommendations for children? Do they know how much time the average student spends online? (Hint, it's a lot more than the AAP's recommendation.) Does your community have agreements in place that cover how and when devices should be used in school? Are your students taught how to manage their own time online?
- Information Literacy. Do your students know how to use search tools and do online research? (This is more than just "Googling," by the way.) Do they know all about Copyright, Fair Use, Public Domain, Creative Commons? Can they explain how Wikipedia works? Can they appropriately cite online sources and do they know how to avoid plagiarism?
- Media Literacy. Does your school teach its students how to critically evaluate media? Do they know how to spot [an] online scam, tell if a website is credible, or evaluate a news source for its veracity?
- Proactive Participation. Are young people taught how to create media? Do they use their devices to blog, make videos and music, create stunning presentations, maybe even code? Do they participate with others, connect with those who have common interests, assert their civic voices?
- Problem Solving. Do students know who to turn to if, and when, problems arise? Are administrators, counselors, teachers and/or parents in your community adequately prepared to deal with potential problems? Have students been equipped with skills to look out for one another online and avoid problems in the first place?

In the end, data encryption, data privacy, student data privacy, and digital citizenship will be under consideration from various perspectives moving forward. For educators and for individuals, it will be important to watch its development because it will have consequences in the technologies we use, mobile or otherwise.

References

Essers, L. (2013, November 20). Berlin court rules Google privacy policy violates data protection law. *PC World*. Retrieved from http://www.pcworld.com/article/2065320/berlin-court-rules-google-privacy-policy-violates-data-protection-law.html

European Commission. (2014). *Factsheet on the "Right to be Forgotten" ruling (C-131/12)*. Retrieved from http://ec.europa.eu/justice/data-protection/files/factsheets/factsheet_data_protection_en.pdf

Evans, M. P., & Saultz, A. (2015, June 10). The opt-out movement is gaining momentum. *Education Week*. Retrieved from http://www.edweek.org/ew/articles/2015/06/10/the-opt-out-movement-is-gaining-momentum.html

Greenwald, G. (2014, October). Why privacy matters. Retrieved from http://www.ted.com/talks/glenn_greenwald_why_privacy_matters/transcript?language=en

Harris, E. (2015, August 6). Connecticut to require all 11th graders to take the SAT. *The New York Times*. Retrieved from http://www.nytimes.com/2015/08/07/nyregion/connecticut-to-require-all-11th-graders-to-take-the-sat.html?_r=0

Hern, A. (2014, May 1). Google stops scanning student emails after California lawsuit. *The Guardian*. Retrieved from https://www.theguardian.com/technology/2014/may/01/google-stops-scanning-student-emails-california-lawsuit

Herold, B. (2014, March 14). Google under fire for data-mining student email messages. *Education Week*. Retrieved from http://www.edweek.org/ew/articles/2014/03/13/26google.h33.html

January, B. (2016). *Information security: Privacy under siege*. Minneapolis, MN: Twenty-first Century Books.

Kharif, O. (2014, May 01). Privacy fears over student data tracking lead to inBloom's shutdown. *Bloomberg*. Retrieved from http://www.bloomberg.com/bw/articles/2014-05-01/inbloom-shuts-down-amid-privacy-fears-over-student-data-tracking

Kharpal, A. (2016, March 29). Apple vs FBI: All you need to know. *CNBC*. Retrieved from http://www.cnbc.com/2016/03/29/apple-vs-fbi-all-you-need-to-know.html

Leong, B. (2015, November 13). Student privacy pledge reaches 200 signatories! *Future of Privacy Forum*. Retrieved from https://fpf.org/2015/11/13/student-privacy-pledge-reaches-200-signatories/

Oliver, J. (2016, March 13). *Last week tonight with John Oliver* [Video file]. Retrieved from http://www.hbo.com/last-week-tonight-with-john-oliver/episodes/3/64-march-13-2016/index.html

Pitas, C. (2013, December 25). Snowden warns of loss of privacy in Christmas message.

London: *Reuters*. Retrieved from http://www.reuters.com/article/us-usa-snowden-privacy-idUSBRE9BO09020131225.

Powles, J. (2014, May 15). What we can salvage from "right to be forgotten" ruling. *Wired*. Retrieved from http://www.ict.pcm.gov.lb/ict/English/NewsDetails.aspx?pageid=5454

Ritvo, D. T., Bavitz, C., Gupta, R., & Oberman, I. (2013, November 14). Privacy and children's data—An overview of the Children's Online Privacy Protection Act and the Family

Educational Rights and Privacy Act. *Berkman Center Research Publication*, *23*. Retrieved from http://papers.ssrn.com/sol3/papers.cfm?abstract_id=2354339

Schneier, B. (2015). *Data and Goliath: The hidden battles to collect your data and control your world*. New York, NY: Norton.

Smith, C. (2016, March 14). Watch: John Oliver perfectly and hilariously explains the FBI's war with Apple. *BGR Media*. Retrieved from http://bgr.com/2016/03/14/apple-fbi-iphone-john-oliver/

Sterling, G. (2013, November 21). German court: Google privacy policy violates consumer rights. *Marketing Land*. Retrieved from http://marketingland.com/german-court-googl e-privacy-policy-violates-consumer-rights-65916

Strauss, V. (2015, May 3). Why the movement to opt out of Common Core tests is a big deal. *The Washington Post*. Retrieved from http://www.washingtonpost.com/blogs/answer-sheet/wp/2015/05/03/why-the-movement-to-opt-out-of-common-core-tests-is-a-big-deal

Student Privacy Pledge. (2015). K-12 school service provider pledge to safeguard student privacy. *Future of Privacy Forum and The Software & Information Industry Association*. Retrieved from https://studentprivacypledge.org/privacy-pledge/.

Vance, A. (2016, April). *Policymaking on education data privacy: Lessons learned*. Alexandria, VA: National Association of State Boards of Education. Retrieved from http://www.nasbe.org/wp-content/uploads/Vance_Lessons-Learned-Final.pdf

Vijayan, J. (2014, April 23). Privacy jitters derail controversial K–12 big data initiative. *Computerworld*.

Retrieved from http://www.computerworld.com/article/2488561/data-privacy/privacy-jitters-de rail-controversial-k-12-big-data-initiative.html

The White House, Office of the Press Secretary. (2015, January 12). *Remarks by the President at the Federal Trade Commission* [Press release]. Retrieved from https://www.whitehouse.gov/the-press-office/2015/01/12/remarks-president-federal-trade-commission.

Zetter, K. (2015, October 8). California now has the nation's best digital privacy law. *Wired*. Retrieved from https://www.wired.com/2015/10/california-now-nations-best-digital-priv acy-law/.

PART II

GLOBAL PERSPECTIVES ON MOBILE TECHNOLOGIES: ONLINE SOCIAL NETWORKS-PRACTICES AND PERCEPTION OF YOUNGSTERS (9–16), THEIR TEACHERS AND PARENTS IN PORTUGAL

The research project that Chapters 5, 6, and 7 of this book are based on is aimed at understanding how Portuguese youth (9–16 years), their teachers and parents, use *online* social networks (OSN); the practices and interactions they develop in that environment; how they perceive the risks, opportunities, and learning on these platforms.

The research objectives were to:

- Understand how young people (9–16 years), their teachers and parents, use the OSN;
- know the practices that they develop in the OSN in areas such as information, publishing, sharing, review, and communication;
- understand the types of links and relationships they establish in the OSN;
- understand how young people, their teachers and parents, perceive risks and opportunities related to the use of OSN;
- Describe actions that young people, their teachers and parents, develop to cope with the risks associated with the use of OSN;

- Know the perceptions of young people, their teachers and parents, on learning in the OSN;
- Know the perceptions of young people and their teachers on the use of OSN by schools for educational purposes.

· 5 ·

USES AND PRACTICES

> The capacity of mobile technologies, as convergent devices offering an ecosystem of connectivity, sociality, and spreadability, calls into play their role in the facilitation of dynamic collaboration, inclusion in civic life, and the ability to coordinate and engage in new forms of civic practice, design, and participation.
>
> —MIHAILIDIS, 2012

Note: Y=Youth (Y), T=Teachers (T), P=Parents (P), and INT= Interviewer (INT)

Introduction

How are online social networks used or may be used in the training of youngsters aged 9–16, for them and with them in formal, non-formal, and informal learning contexts taking into account the uses, practices, inter-relationships, and perceptions of these young people, their teachers and parents, regarding these digital platforms? This general question led to sub-questions, which we organized in five areas:

- Uses—Young people (aged 9–16), their teachers and parents use OSN? Which ones? How often?

- Practices—What are the most common practices—information, publication, sharing, review, and communication—of youngsters, their teachers, and parents in the OSN?
- Relationships—What interrelations establish youngsters, their teachers, and parents in the OSN?
- Risk perception—What risk perceptions have teachers and parents regarding the use of OSN for their pupils/students? Youth, their teachers and parents have strategies to avoid the risks associated with the use of OSN?
- Learning—Youth, teachers, and parents consider or not the OSN learning spaces? Youth and teachers consider using the OSN at school for educational purposes?

Data Instrument

The research followed the *design* of mixed methods. The methodological option was within the sequential explanatory approach, that is, first is the collection of quantitative data, then analysis and organization of the qualitative phase depending on the results: the initial qualitative data are explained with qualitative data (Creswell, 2014).

The questionnaire for young people (9–16 years) had 32 questions, one with Likert scale and nine with intensity scales. It focused on customs and practices on the Internet in general and in the OSN in particular; the perception in terms of technical monitoring and contents; the relationship with peers, parents, and teachers through the OSN; the perception of risk and opportunities; and the perception in learning terms. They were also asked if they agreed with the use of OSN for school and educational purposes.

The questionnaire for teachers had 25 questions, seven with intensity scales, focusing on uses and practices in OSN, perception of uses by the students, the relationship with students and parents through the OSN, the pedagogical potential of OSN, risk perception that students are subject to, and actions taken to deal with any potential risks (those adopted for themselves and those intended for students). Only teachers who taught the classes of surveyed students answered it.

The questionnaire had 22 questions for IT, with five intensity scales. It focused on the uses and practices in OSN, perception of uses from their students, the relationship with students and their teachers through the OSN, perception of risks that the students are subject to in the OSN, and actions taken to deal with any potential risks. Only parents that had allowed their students to answer the questionnaire answered it.

From the analysis of data in Statistical Package for Social Sciences (SPSS), we organized three interview guides (one for each sub-sample), which aimed to better understand the results, proceeding to their interpretation with the support of previously surveyed subjects, using interviews. The interview guides were organized into five key blocks from the relationship with the OSN, the relationship with the contact list in the OSN (including relationships between young people, teachers, and parents), practices in the OSN, the perception of risk and opportunities, the relationship between OSN and school. We proceeded further to the interview legitimating; we addressed the practices on the Internet and requested suggestions for research.

The Surveyed Sample

We produced and validated qualitatively three questionnaires with teachers, young people, parents, and experts in Portugal, France, and Italy. The questionnaires were applied in 11 Portuguese schools between January and April 2013 to a convenience sample of 549 young people, 276 males (50%) and 273 females (50%) from 10 to 18 years. About 4% were between 16 and 18 years (12 individuals) and 57% were up to 12. They attended the second cycle (41%) and the third cycle of Basic Education or the 10th grade (59%), and the largest group was the sixth grade, followed by the groups of ninth and seventh grade.

Of the 266 parents who responded, 208 (78%) were female and 58 (22%) were male. As expected, taking into account the age of the young people surveyed, most parents (93%) were between 31 and 50 years. One of the parents did not give their age. A third of parents (33%) had completed secondary education, but more than a quarter of them (29%) had not completed compulsory education, which is the 12th grade, and only 10% had completed the first or second cycle. One did not respond and another stated he could not read or write. In terms of employment status, 72% of parents were full-time workers. About 11% worked part-time (6%) or were housewives (5%). The unemployed accounted for 14%.

A total of 150 teachers answered the questionnaire, 65% female. The majority (57%) was between 41 and 50 years, 22% was 51 or more, and 21% up to 40 years. Of the 150 teachers, 72% of teachers were graduates, 16% MA, and only 1% PhD. Already 6% were college graduates and 5% were postgraduates. On the whole, respondent teachers taught all subjects that were taught in the second and third cycles. We organized teachers in groups. Most teachers taught more than one subject, in a total of 197 subjects, distributed as follows: Languages (64), Exact Sciences (49), Art (37), Social Sciences and Humanities (22), and others—Moral Catholic Religious Education, Information and Communication Technology, Special Education, and Citizenship Education—(25).

In the qualitative phase we interviewed 142 of the 549 young people surveyed in the quantitative phase of the study, 52% male, 48% female; 60% were up to 13 years; 17% attended the second cycle and 83% the third cycle and secondary school. Twenty parents aged between 30 and 51 years, 14 female, were interviewed. Of the 20 teachers surveyed aged 42 and 64, 13 were female and taught Languages (10 teachers), Exact Sciences (four teachers), Art (two teachers), Social Sciences (three teachers), and Others (an ICT teacher).

Study Restrictions

A set of restrictions, in both phases of the investigation, should be considered when interpreting and eventual generalization of the results found: 1) is a non-experimental study, which does not establish causal relationships; 2) questionnaires were generated and validated by the researcher, so they had not been applied earlier; 3) the data were collected through a convenience sample and may not be generalized to other contexts; 4) young people who participated in the focus interviews, just as with individually interviewed teachers and parents, were those who made themselves available for this purpose (or were authorized by parents, in the case of young people), so the qualitative results could be different if the respondents were other than these.

Results

The results show that most young people, teachers, and parents use online social networks, but tend to ignore each other in those spaces. They look for information about the present and participate in social networks, though more as reproducers than as content producers.

Most teachers recognize the pedagogical potential of online social networks, but only a minority use them for this purpose. Teachers and parents are more focused on the risks than on the opportunities and fear the consequences they may be subject to as users, especially where young people are concerned. Yet, there is a belief among young people, teachers, and parents that there is potential for learning with online social networks. At the same time a concern exists that social networks co-exist in isolation, ie. formal and non-formal, which poses a question as to whether there can be an educational benefit to using OSNs.

We identified shortcomings regarding the use of online social networks for educational purposes and active citizenship. But these limitations can be overcome if young people, teachers, and parents gradually become literate in the media, which implies the urgent and effective integration of Media Education in all learning contexts.

Recommendations

The results of this research suggest two guidelines in terms of continuity of the relationship between OSN and learning: 1) it is clear that there is a need to achieve Media Education, which "consists in teaching about media besides teaching resorting to the media" (Hobbs, 1998, p. 10), which implies being aware of the representations of the actors involved, and the actual practices and shortcomings observed at the level of their knowledge and know-how; 2) it is clear that there is a need to develop projects with children of pre-school and first cycle because the young start contact with the OSN before 10 years of age. Only a quarter of young people at that age did not have a profile on an OSN, and 40% of 499 users surveyed had their profile at 8 years or before. The entrance into the OSN may happen even earlier, as noted by a parent we interviewed:

> And even my other son, at six, he memorizes what we do. If I give him freedom, tomorrow he is on the network. He is able to do this. Obviously he does not have that freedom. They seize it and, if they have freedom proceed. (P6)

This statement is consistent with the results of a study on the use of technologies and media, including the Internet, by European children from zero to eight years old within family. After in-depth interviews with 70 children and at least one parent, the study concluded that "children learn watching others, especially parents and other family members," including older siblings, as well as grandparents, cousins, uncles and even neighbors. However, "in most cases, parents do not seem to be aware that children reproduce their behavior," and some even discovered during the interview that their children knew their passwords (Chaudron, 2015, p. 14).

In addition, the research has focused on children and young people from the age of 9, because only 12% of about 1200 identified by the EU Kids Online in Europe included children under the age of 7 years, while only 20% included the perspectives of teachers and 13% the perspectives of parents. On the other hand, the research has focused on Internet use by children and young people and not learning with the media about the media and through the media (O'Neill & Staksrud, 2014). Now the projects should not only focus on Internet access and web use, but also on the role that parents and teachers are, or are not, having in terms of preparation for the active use of technologies (Ólafsson, Livingstone, & Haddon, 2013).

We are currently developing a 3-year project that includes training of pre-school teachers and first cycle in Media Education, followed by monitoring and support in carrying out educational activities in the classroom, using digital technologies. This monitoring will last for 2 years and also includes students and their parents. We chose this project because the early years are critical to the development of abilities and skills in social media, while it is necessary to change the focus of the research from risk to opportunities. According to Jenkins (2009), this project also aims to contribute to the challenge of ensuring that all benefit from the expansion of the media landscape. This can only be possible reducing the gap between the new cultural elite who have access to the media and participate effectively in and through them, and the cultural underclass that, even having access, does not participate because it does not know how to.

Web 2.0, Social Media, and Online Social Networks (OSNs)

Web 2.0 differs "from social media, because it does not refer to any particular application," but appoints the new uses of the Internet "as a platform in which content and functionalities evolve continuously, thanks to the activity of users" (Endrizzi, 2012, p. 3). It is a social web model characterized by multiple possibilities of interaction by users, and has led to the emergence of a new information distribution topology (Terranova & Donovan, 2013). It is characterized by being easy to use, it promotes socialization, and it enables the production and free publication of content in different formats, from photographs, videos, or text (Lovink, 2011) on different platforms that are either publicly accessible or accessible by signing in. Kaplan and Haenlein (2010) ranked it into the following groups: 1) collaborative projects such as

Wikipedia; 2) blogs (e.g., Blogger); 3) virtual games (e.g., Counter Strike); 4) virtual worlds (e.g., Second Life); 5) social networking sites (e.g., Facebook); 6) microblogs (e.g., Twitter); and 7) content communities (e.g., YouTube).

Social media are the Web 2.0 platforms that enable us to publish, share, play, rediscover people, buy online, and locate places, among other features. Cavazza (2015) organized them according to four types of use: publication (e.g., Wikipedia, Blogger); sharing (e.g., YouTube, Vimeo); conversation (e.g., reddit, Yammer); and networking (e.g., LinkedIn, tinder).

Facebook and Twitter are like hubs because they "allow users to fulfill each of the four main social usages, but they also stand as relay of information" (Cavazza, 2015), while mobile applications such as WeChat, Messenger, Hangouts, WhatsApp, Viber, Kik, and Snapchat, which began as mere communication tools, have become sophisticated applications that enable actions such as calling a taxi, sending money, playing, or reading full articles; that is, they try to be "the digital Swiss knife in your pocket" (Cavazza, 2015, p. 10).

The diversity of sites and applications offered on the computer or on smartphones or tablets is large, perhaps too big for all to remain under the umbrella concept of Social Media. It matters to distinguish them from each other, to be able to study them, avoiding one of the two more common attitudes: admiring social media or hating social media (Stenger & Coutant, 2011). Thus, we considered the online social networks (OSN) such as Facebook as a special case within social media. We also did it taking into account our main focus: the youngsters, who quickly adapted digital tools to their interests, particularly in terms of entertainment and use of OSN (Underwood, Parker, & Stone, 2013).

Although researchers agree about the importance of studying the uses and effects of OSN, there is no consensus in the literature regarding a precise definition, or on the criteria that must be met by certain platforms to be considered an OSN (Panek, Nardis, & Konrath, 2013). We have chosen, however, to consider the definition proposed by UNESCO (2011), according to which an "online social network" is a group of people connected to the Internet who share common interests and activities, being able to publish profiles and information about themselves. A profile on an OSN includes: photographs, and one's own (or others') videos; cultural preferences (e.g., movies, music, books); one's contact list; the groups to which he belongs; and personal preferences, places/locations, and publications, organized in reverse chronological order. It may represent a person, but it also represents: "cultural objects (movies, music); consumption (products); places (cities, institutions, associations); social movements (parties, representative groups, ideologies); and other nodes not human" (Rosado & Tomé, 2015).

The profile can be public or semi-public, and you can use it to generate a list of contacts with which a connection is established; you can browse the publications of those on the list and the publications of people that they are associated with (boyd & Ellison, 2007). Besides browsing the publications, following Cavazza (2015), the profile enables one to participate through publication and content-sharing (text, photos, videos, links), conversation with contacts—synchronously or asynchronously—and networking, belonging to micronetworks within the OSN (e.g., a person is on LinkedIn, not in connection with all but only with their associates). Restrictions on "time (simultaneity) and space (proximity) can be overcome in these new spaces of socialization," creating/allowing/enabling interactions among people who are "far apart geographically" and in different time zones (Rosado & Tomé, 2015, p. 15).

OSN use is mediated by technological factors that enhance four historical characteristics of content generated by citizens (boyd, 2015): 1) Persistence (published content persists over time and can be viewed or reviewed beyond the time they occurred); 2) Visibility (certain messages are as visible as the audience is large); 3) Replicability (sharing a message at the touch of a button); 4) Searchability (search engines, in general, or of the OSN), easily enable one to find content.

The presence on the OSN aims to socialize, so users (re)present themselves to others, the "networked publics" (boyd, 2015), minimizing the boundary between public and private, as the individual builds his profile and is exposed on the OSN, revealing fragments of his intimacy, waiting to see them validated by other users in a process that Tisseron (2011) called "extimacy" ("*extimité*").

The OSN minimize the boundary between the real and the virtual, so it is essential to develop control strategies for one's representation in the OSN, since the construction of this representation is a collective rather than an individual process (Enli & Thumin, 2012). The contacts of the OSN validate, or not, what the subject reveals about himself, and participate in the representation of this subject through the content they publish.

Although the OSN have a global scope, the participation of individuals is mediated by the political, economic, social, cultural, and technological contexts to which they belong. The OSN change the nature of the groups, social formations, and power, and the way of assigning meanings to *media* contents, so that they transform society and popular culture, in particular (Gee, 2010). In the meantime, OSN use is growing exponentially.

Facebook is the OSN that currently has the most users, about 1490 million worldwide (Statista, 2015). We can consider it a "parallel network" within the Internet as it competes in terms of the amount of access with the largest search engine, Google; Facebook had 207 million unique viewers in December 2014 in the United States, behind Yahoo Sites with 216 million and Google with 238 million (ComScore, 2015, p. 6).

In the United States of America (USA), 89% of young people aged 13–17 years use at least one OSN. Facebook is the most used (71%) (Pew Research, 2015). Use is also high among adults, with Facebook leading. It is used by 81% of the American adult population who use OSN, and in December 2014, users spent 230 billion minutes using it, which corresponds to 18 times the usage time of the second-most-used OSN in the US, Instagram (ComScore, 2015, p. 10).

A European study developed in seven European countries, including Portugal, reveals that, on average, 68% of young people (9–16 years) use OSN, with Facebook leading, as 61% stated that they have a profile on that OSN (Mascheroni & Cuman, 2014). Still in Europe, taking into account the entire population (users and non-Internet users) from ages 16 to 74, 46% use OSN, with significant differences between the age groups 16–24 years (85% use), 25–54 (52%), and 55–74 years (16%) ("Individuals Using the Internet," 2015).

In Portugal, 64% of youth aged 10–18 use OSN on a daily basis, and Facebook is the preferred OSN by 86% of users (Lopes, 2015). According to the project Net Children Go Mobile, 76% of Portuguese aged 9–16 have a profile on an OSN, and 72% have a Facebook profile (Mascheroni & Cuman, 2014, p. 17). Among the Portuguese population between the ages of 16 and 74, 47% use OSN, with considerable differences among the age groups 16–24 (94% use), 25–54 (54% use), and 55–74 (13% use) (Obercom, 2015).

Access to the OSN is done progressively through mobile devices, so we see today a "convergence between mobile and social networks" (Endrizzi, 2012, p. 5). Smartphone users are involved in more online activities, such as visiting OSN profiles, which is related to five advantages of this equipment over the laptop (Mascheroni & Cuman, 2014): 1) Convenience (smartphone is small, portable); 2) Privacy (is of private use); 3) Publication (enables the publication and quick sharing of photos using apps); 4) Contact (it enables being permanently attached to contact groups, even with those who are not on OSN or who do not have Internet access (through SMS, Short Message Service).

Being aware that a research work in the Media Education area implies, "Start by taking consciousness or helping others be aware of the relationship that each one has with media" (CLEMI, 2013, p. 8), we started the project "Social Networks and New Media Literacy" (2012–2015), characterizing the online life of young people, their teachers and parents, their practices, their interrelationships and languages.

Results

Although the total sample was 549 young people, 150 of their teachers, and 267 of their parents (see sample characterization in the Introduction of this section), on issues concerning the use of OSN or perceptions of OSN, we only considered the feedback from users: 499 young people, 110 teachers, and 172 parents.

Uses of OSN

There are more young people with at least one profile on OSN (91%) than teachers (73%) and parents (65%). Among users of OSN, Facebook is the most popular, with similar percentages among the three groups (Figure 1).

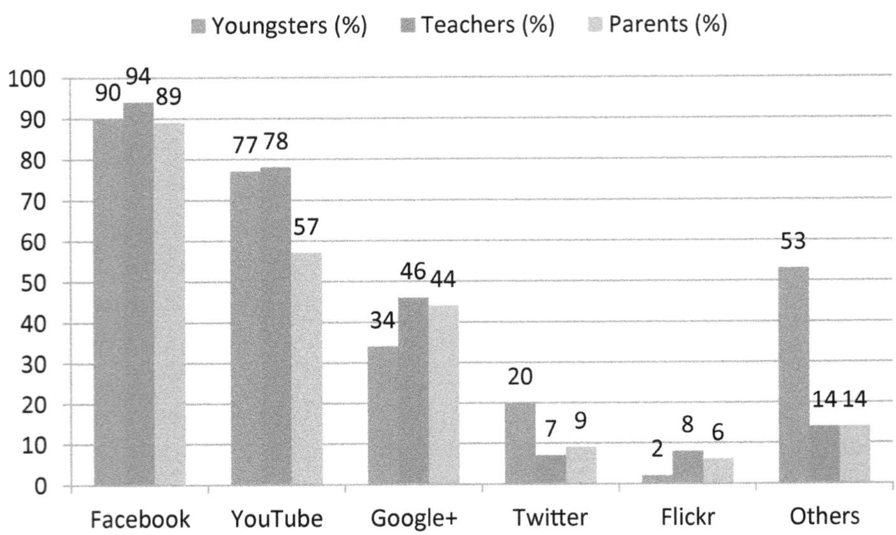

Figure 1. Use of OSN by young people, teachers, and parents (%) N = 499 (youngsters); 110 (teachers); 172 (parents).

YouTube was most used by teachers and young people rather than by parents, while Google+ was used more by adults than teens. Twitter (20%) was more for young people and Flickr wsa barely used. For "Others," the percentages were low among teachers and parents, but show that more than half of young people used other OSN, such as Tumblr (12% of total) and ASKfm (8% of total).

Among parents, there were more males (76%) using OSN than females (61%). There were even more parents in urban areas using OSN than in rural areas (73% and 57%, respectively). In the subsample teachers, we did not find significant differences.

Among young people there were more boys (94%) than girls (88%) using OSN. The use increases as one progresses within the young age group (3/4 at the age of 10; around 90% between the ages of 11 and 13; and 100% boys and 95% girls at the ages of 14 and 15).

A third of parents, 10% of young people (50 individuals, 17 boys and 33 girls), and 25% of teachers surveyed did not use OSN. Youngsters assigned five reasons not to have a profile on OSN: no parents' approval; they stopped using; no Internet access; detachment from OSN ("When I want to talk to other people I can use other things. I do not see much importance to be [on Facebook]"; José, 13); or they use one of their parents' profiles ("I use the profile of my mother. I am the one who uses most, but there under the name of my mother and her data"; Carlota, 12).

The three teachers interviewed who had no profile on OSN: expressed disinterest in OSN ("I've never used. I am outdated ['*demodée*']. Some call me abnormal, but I fit... networks, to me, are meaningless"; T31); they no longer use it because they are close to retirement; or they have no need of its use, because they use other ways to communicate:

> What I need to do, I do by email....To communicate with other colleagues I do not need Facebook, but I use other tools....If I feel the need, as a teacher, to use this platform I would use it as an e-mail or as I use a web page. As I do not feel the need, I do not use. I never felt curious [about] having a Facebook account because I would not use it for [anything]. (T2)

All parents who agreed to be interviewed had a profile on an OSN, which was a limit to interpret data and can be read as an indicator of the parents' detachment, of those who do not use OSN.

Facebook: Between Daily and Occasional Use

Among users of OSN, Facebook is the most used, being mentioned by 91% of respondents, of which 39% used it every day and 18% did so between 4 and 6 days a week. The usage frequency had a common pattern. We identify two trends: 1) the daily use by 39% of young people and about one third of teachers and parents; 2) the sporadic use, less than one day a week, reported by about 20% of teachers and parents and 12% of young people. Teachers are the ones who claimed to use it 5 days a week (11% vs 6% of parents and 4% of young people), which may be related to non-use on the weekend, as was reported by some teachers in interviews. Among users of OSN, there were only 10% of young people, 11% of parents, and 6% of teachers who said that they did not use Facebook.

Young People Do Not Give Up

If Facebook is the OSN with more users, this does not mean it is the most used ("Before I used Facebook for many things….I played almost every day on FB, then I never played"; Francisco, 15). But if young people start using a new OSN with some frequency and in a group, some of them not being visible to parents, the loyalty to Facebook remains, which we documented.

> Y8—We do not abandon, not completely. I do not have classmates logging out of Facebook, all at once. What can happen is that instead of accessing every day, they will do it once or twice a week. They do not leave Facebook straight away.
> Y3—The problem is that it has become a habit to surf Facebook. It is already an addiction.
> Y1—People just go there to see what is happening. They do not stay there.
> Y8—Yes, they do not stay there for a long time.
> Y2—They see and see, and keep going...
> Y6—It's not as interesting as it was. That is why we started surfing other networks.
> Y8—Yes, because we spent a lot of time...for example, on Facebook, it is no longer funny, there are no more secrets.

Those who give up Facebook go back to it: "I have some friends who live abroad or far away" and "to speak to them is much easier with Facebook" (Pedro, 16). But the site of the OSN remains open "only to check what is happening" (Catarina, eighth grade, 13 years), since, although it is not seen as "inherently good," it is "where we find everyone" that young people know; it is one of the few public places where "teens can easily join with their peers" (boyd, 2015, p. 39):

Y5—As most friends have Facebook, we...we talk there.
Y7—You also can play...
Y6—I only play and talk, nothing else.
Y2—Yeah, me too.

The innovation of other OSN captures young people's attention, as well as their easy access and the ability to have different activities in each of them: "Instagram and Tumblr are more to publish photos and stuff. Facebook is more to see moods and to talk" (Luis, 12); or, taking into account their features: "On YouTube I watch videos. On ASKfm I even share my Facebook profile, to ask questions" (Fernanda, 12). The diversity is also related to the Internet access equipment:

> I used Facebook in August nearly two years ago and I used it a lot until I got information off of Instagram, which was very cool. I created an [Instagram] account and posted more. As I have Instagram apps on my mobile phone, I log in [there] more often than Facebook. It is much faster and is almost the same thing. (Guilherme, 13)

Young people admit that "Snapchat has also gained popularity" (Carlos, 16) and "WhatsApp is also one of the options that people are using more" (Mariana, 15), as "with new technologies, people access the links more easily and surf other networks" (Alexandra, 13). "Children simultaneously use various services, each enabling specific practices and targeted at a specific audience" (Mascheroni & Cuman, 2014, p. 15). They often reach the computer and open all OSN where they have a profile, browsing among them: "When I am on Facebook I open new tabs and go to ASKfm and YouTube" (Gonçalo, 12). Diversification is not just a youth practice, as we found out after organizing the respondents in groups according to the number of OSN. Young people used, on average, 2.8 OSN, teachers 2.5, and parents 2.3.

The adults were the group who most used only one OSN (32% of parents and 20% of teachers), the residual percentage among students (7%). Young people were the ones that used two OSN (37% versus 33% of teachers and 29% of parents), but teachers led in the use of three OSN (36%, against 30% of adolescents and 26% of parents). The diversity comes at four OSN, particularly among young people (19%), since adults are already a minority (around 6%). Users of five or six OSN do not exceed 7% in each of the sub-samples.

Already 238 young people said they use other OSN beyond the indicators in the questionnaire, but only 199 completed the answer and of those only 106 related OSN in the case of Tumblr (58 users), ASKfm (41 users), hi5 (eight users), DeviantArt, Instagram, MySpace, Tuenti, VK (VKontakte), Stickam (now closed/no longer available), PlayStation Network, Badoo, Spring.me (formerly Formspring), and Tagged (one to three users). The others had difficulty distinguishing OSN from other applications on the Internet, which is understandable since in a culture of media convergence (Jenkins, 2006), it is often observed that teenagers, especially, no longer distinguish the different types of platforms or the diversity of contents (LeDeuff, 2011). Teachers refer to only hi5 and Pinterest as other OSN used. Among parents, other OSN assume residual values (LinkedIn, Pinterest, Instagram, Foursquare, and Tumblr).

Youngsters Start Using OSB Before 13

Nine out of 10 students stated they have created their profile before the age of 13. Some admitted to having created their first profile at 6 years (12% of young people age 10) or 7 years (14% of young people age 11). Only a quarter of the 14s and half of 15s (45 individuals, 9%) will have created their profile at 13 or later (Figure 2).

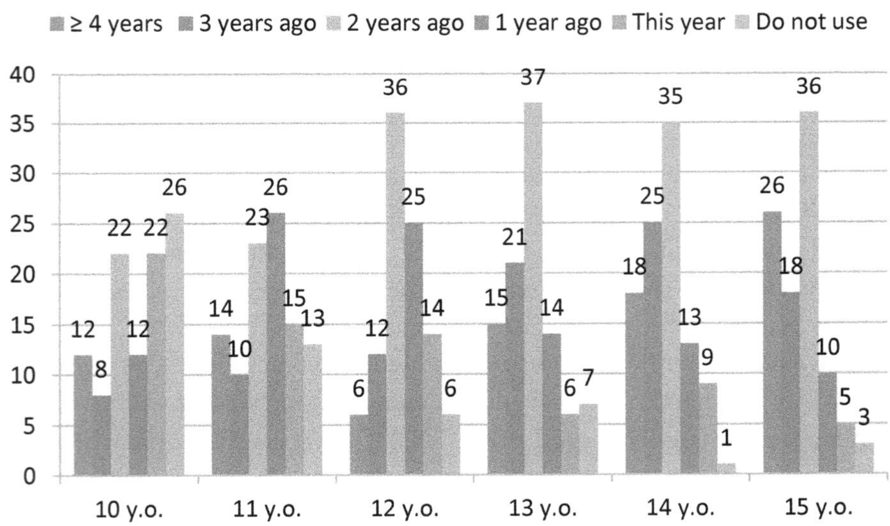

Figure 2. Number of years that young people created the first profile in an OSN, by age group (%), N = 487.

The issue was raised because in the United States, the minimum age to join Facebook is 13 years, as followed by commercial websites and others as a way to not conflict with federal law (Children's Online Privacy Protection Act), which requires previous parental approval for children less than 13 years. In the "Statement of Rights and Responsibilities" (2015), Facebook stated "You will not use Facebook if you are under 13" (para. 4.5).

Age is not a limiting issue of access to OSN. The data collected shows that input in OSN is regarded as normal by young people: at 10 years, a quarter of them have no profile in OSN; the percentage decreases to 13% at 11 years and 6% at 12 years, having residual values at 14 and 15 years. The interviews confirmed the results:

> INT—91% of students got into the networks before 13 years. From what you know of your peers is it really like that, or not?
> Y7—Yes, it is like that with everybody.
> Y1—When I created mine I already had colleagues who had one for many years already.
> Y3—I know a person who was in the 2nd grade and already had Facebook.
> Y5—They create [accounts] to play games.
> Y6—I created [an account] maybe when I was in the 5th grade and there were people who already had.

Young people often say that they had their parents' agreement to create their profile and, in some cases, it was even their parents who created it. This is now a common practice, preventing the students from creating their profiles themselves, and at the same time parents can be associated with these profiles and do some monitoring on student activity in the OSN. According to Morduchowicz (2014), parents do not forbid children of 8 or 9 from crossing the street, but they create conditions to monitor them, so they may act similarly in the OSN. Something that exists among young people surveyed:

> Y2—My father [created] my Facebook [account] and I did not know I had it.
> Y6—I came home every day saying I wanted to have Facebook. My father, in the summer holiday, created my profile. I begged him so much. He told me to go there and then he did it in front of me.
> Y1—It was my mother who helped me.

Three quarters (79%) of parents said their children have a profile on an OSN. Half of parents whose students are 10 stated that these children have a profile on OSN, raising the percentage for three quarters at 11–12, exceeding 90% from 13 years. All parents of students age 15 confirmed the presence of these young people on the OSN.

All parents interviewed reported that their students are OSN users and began their involvement before 13. "Do not have the slightest doubt of that.

My youngest has [had] Facebook from the third or fourth grade. So she was about 8. And she created the account on her own" (P6). In 10 of the cases where the parents themselves helped to create the profile, six of the students requested permission from their parents in this way:

> Before creating the profile, my son asked me if he could do it or not. There was a prior conversation about cautions to take and then he created it. Still today he asks me or [his] mother if he can do this or that on the network. (P5)

Four parents did not expressly authorize their students to create profiles, but interacted with them on the OSN or tried to understand the actions they developed there. "He accesses the online platform in the living room where we all are. I see what he shares, [the] friends he talks to....The games...I do not control very well" (P7). But they cannot follow them on all OSN: "He is on Facebook and I think he is on Twitter too. I think he created an account there" (P15).

Only 19% of parents ensured that their children do not use OSN, 2% did not answer, and one (0.4%) stated that they didn't know. Parents who did not confirm the existence of a profile by their child presented as justification the prohibition or the use of other people's profiles:

- "Because [the child] is not old enough" (28 parents),
- 'I do not want [them to have a profile]" (8 parents), or
- "Other" (9 parents).

Among the "other" reasons, it was mentioned that the student used their father's or mother's profile (two cases), "is not interested in it yet" (3 parents), "does not need a profile," "our decision," "because she has physical disabilities and is followed by me," or because "they just deleted it" (one parent).

The results are according to a study that surveyed 1007 parents of young people (10–14): 50% confirmed that their students had a profile on OSN and 36% said that the children created the profile before 13. Among parents whose children had Facebook profiles, 72% said the children had created the profile before 13 years. For children from 10 to 12, parents had helped in the profile creation in two thirds of cases. The authors concluded that the legislation was ineffective, and that there was a need to create other mechanisms that benefit and support children, youth, and parents regarding OSN use (boyd, Hargittai, Schultz, & Palfrey, 2011).

Most teachers also had the perception that students use OSN, including children under 13, since 62% thought that all students, or most of them, have profiles on OSN and 19% said that it's half or less than half, while only 17% said they do not know if students use OSN. When asked about the students' preferred OSN, all teachers who answered the question (81%) claimed it was Facebook, but they made a distinction between students who have Facebook and those who use it:

> I admit they all can be on the network, but I do not believe that all use it. I work with teenagers from seventh to ninth grade and I do not believe they want to be as visible. They do not want to expose. They have Facebook because it is socially recommended, but use it, I have my doubts. (T10)

Teachers referred to diversity in the use of OSN by students ("Facebook is their favorite network, even knowing they use others like Twitter, because there are many"; T13), as young people associate other OSN relative to the use of Facebook:

> There are many kids in our school who use ASKfm, but they are at the same time on Facebook and ASKfm. That is, they share on Facebook situations they post in ASKfm. They are in several networks at the same time....Twitter belongs to the previous generation. Not this generation. They prefer faster networks requiring fewer words. (T7)

Concluding, about 90% of young people are on OSN and nine in 10 have created their profile before the age of 13. There are cases where this happened while still attending primary school. This often happens with the help of family members, including parents, it being recognized by the majority of parents and teachers that young people, even the younger ones, use OSN.

Practices on Social Networks

We compared data of youngsters, their teachers, and their parents for five practices in OSN: 1) news consumption; 2) publication; 3) sharing; 4) comment; 5) communication.

OSN as a Source of News, But Doubtful Credibility

Teachers (89%) were the group who most found current news stories they read, watch, or hear, in the OSN, followed by parents (84%) and young people (77%), and the information consumption was significant in the three sub-samples (Figure 3).

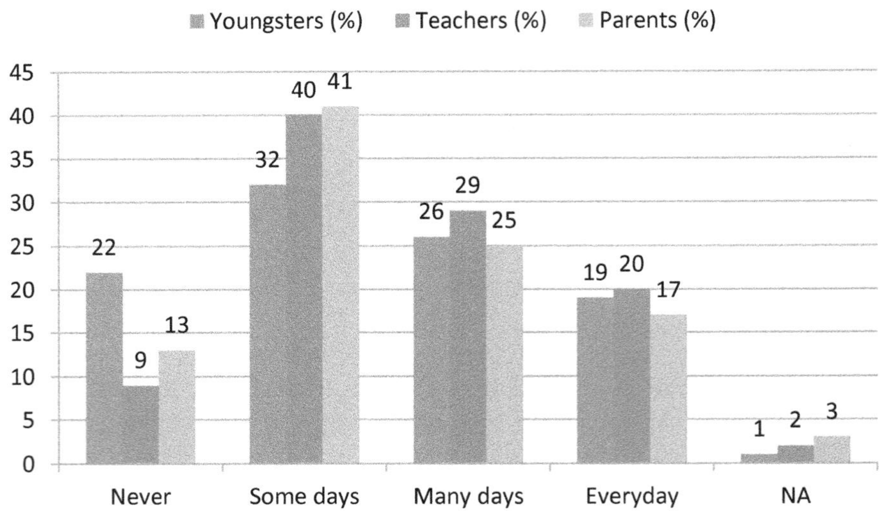

Figure 3. Access to news in the OSN by young people, teachers, and parents (%) N = 499 (youth); 110 (teachers); 172 (parents).

Almost half of teachers, parents, and youngsters stated that they were informed every day (about 20%) or many days (almost 30%) through OSN. Consumption was sporadic for about 40% of teachers and parents as well as 32% of young people.

Regarding young people, the data match with the results of a Canadian study (Steeves, 2014) that surveyed 5436 students from fourth to 11th grades, and concluded that 78% of students used the Internet to find information about news and 49% followed the news and current events. Our results are, however, higher than the results of a European study (Mascheroni & Cuman, 2014), according to which 14% of young people ages 9–16 read or watched news on the Internet. Considering only the young Portuguese respondents in this study, only 10% reported reading or watching news on the Internet, it being a more common activity among boys from 13 to 16 years (Simões, Ponte, Ferreira, Doretto, & Azevedo, 2014).

Among adults, the results we found show similar trends to those found in a European study that surveyed nearly 8,000 consumers of news over 18 years, 1,049 of them in Portugal, and concluded that 86% of the surveyed consumed news online. The study puts the Portuguese among the most active users of OSN to be informed by those sites, since 61% admit doing so, being only surpassed by the Turks (67%, but Turkey only counts its urban population), and ahead of the Polish (52%), Dutch (43%), Czech (41%), and Austrian (38%). Among those who surf the OSN to get informed, Facebook (67%) was the OSN most said, then YouTube (26%), Google+ (18%), and Twitter (7%). The Portuguese are still the ones who declare that they share the news on OSN (Fletcher, Radcliffe, Levy, Nielsen, & Newman, 2015).

Our results match with a Portuguese study that surveyed 1035 individuals (16–74), of whom 67% were Internet users. Among users, 89% said they checked online news, and 55% used it at least once a day. Individuals who most reported that they checked online news belonged to the age group 25–34 and were mostly the better schooled (Obercom, 2015). Regarding our sample, the majority of parents and teachers were situated in the range from 36 to 50, but teachers were among the better schooled, and of parents, only 13% of individuals 35–44 and 14% of 45–54 did not check online news.

News Consumption Varies With Age and Gender

The use of the Internet to keep up with news and current events varies depending on age, because if 49% did so, the percentage increases to 28% among young people in the fourth grade, and to 65% among those who attended the 11th grade (Steeves, 2014). Also among the young people, age influences the consumption of news in the OSN. It increases as we progress in the age group (Figure 4).

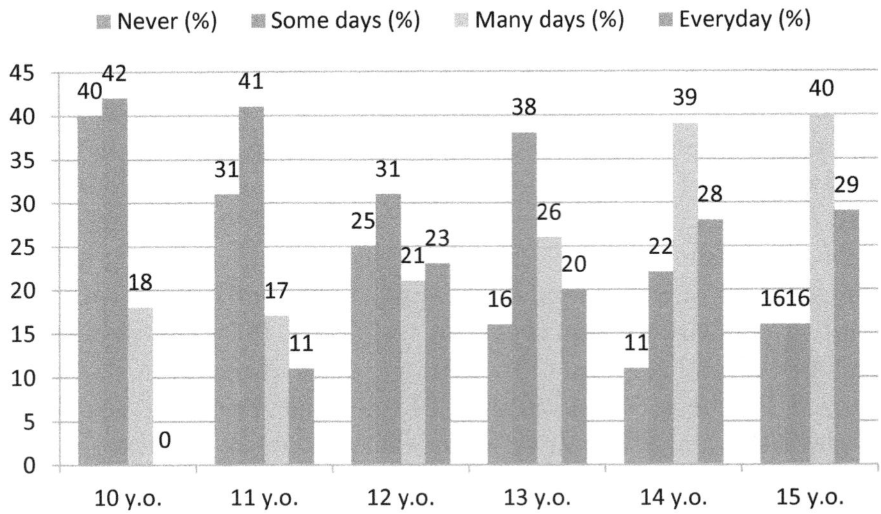

Figure 4. News stories consumption in the OSN by young people, by age (%) N = 483.

At age 10, already 60% declared they got informed on the OSN, although consumption was mainly sporadic (42%) and not even one in this category claimed they got informed there every day. From 10 years, the percentage of those who got informed on the OSN and those who do it more often tends to increase. Young people of 12 were divided into four similar groups (a quarter did not consume and each of the other quarters got informed with different frequency), while from 13 years, more than 80% declared that they got informed on the OSN. At 14 and 15, almost 70% of young people declared that they got informed on the OSN many days or every day.

We organized young people into three groups: 1) those that often got informed on the OSN, tendentiously the group of 14 and 15 ("Often, the news we watch is no longer on television, newspapers, or magazines, but it is on the Internet"; Catarina, 14); 2) those who do "very rarely" (Inês, 12); 3) those who said they did not get informed on the network ("I do not load it"; Joana, 12), tendentiously in groups of 10 to 12 years.

Young people consume news that they find on the OSN "when it is interesting" (Beatriz, 12), "very showy" (Diana, 11), "when it is worth it" (Miguel, 13), "if I am curious" (Mariana, 13), or if it is "any shocking thing" (Vanessa, 14). The subjects that interest them are related to disasters, catastrophes, or accidents (e.g., a train crash in Spain, forest fires, the death of students during an academic party), celebrity deaths (e.g., Paul Walker, Mandela, Eusébio).

The concern with celebrities (e.g., Justin Bieber, One Direction) is more common among girls, while boys focus on sports (e.g., transfers of players, goals, results). The results are consistent with the results obtained by Steeves (2014), which found the biggest differences in the most followed news by children on topics such as sports (63% of boys versus 31% of girls) and entertainment and celebrities (53% of girls compared to 32% of boys). This gender issue is not unique to young people. Among the Portuguese age 15 or more and in terms of the topics of the news considered more important by gender, the main differences are found in sports (important to 82% of men and 18% women), art and culture (39% of men and 61% women), and entertainment and celebrities (40% of men and 60% of women) (Obercom, 2015, p. 16).

Techniques to Verify the Accuracy of News

Despite consuming news on the OSN, young people are suspicious of its veracity, because "some are pure lies" (Gonçalo, 11), or "most of what you see on Facebook is fake" (Francisco, 15), since "nowadays it is possible to create a page to imitate, for example, TVI and SIC [mainstream Portuguese TV channels], and invent things" (Beatriz, 12). Here is an example:

> Y7—A few days ago it was said they found the mortal remains of Maddie McCann [a 4-year-old British girl who disappeared in Portugal in 2007]. That is false. Because then I surfed the SAPO [Portuguese search engine] for the current news, and there was not any of that in the news.
> Y2—And another time they published that the one who plays Hancock, the actor...
> Y1—Will Smith.
> Y2—Yes. They were saying that he had died during the recording of Hancock 2. And besides he did not die, the Hancock 2 is not even being shot.

The statements show that between 2010 and 2014 young people improved "critical skills such as comparing different websites to assess the reliability of information" (Mascheroni & Cuman, 2014, p. 19), as evidenced by the information confirmation strategies referred to by young people, i.e., whether they are more or less effective:

a) Confirming on TV—Although they are aware that there "have emerged stories on television and there were problems with that" (Francisco, 16), it is TV they trust more ("I watch [news] on television because it is more true. We see people talking, it is not just anyone writing"; João, 14), or because "the Internet is much faster, but television is more reliable; [why] it is shown there is because it happened" (David, 15).

b) Scanning the mainstream media websites, to which they grant credibility— "Sometimes, to better see the news, we check 'Bola TV'" (Rodrigo, 12). "If it is a news [story] in the Daily News, normally [it] is true" (Inês, 15).
c) Searching on Google—"I go to Google, type it, and try to figure [it] out" (Bruna, 12); "When Paul Walker died, I saw on Facebook and I could not believe it. I thought it was a lie. But then I checked in Google and confirmed it was really true. And this has happened in other situations. I had to check in Google to confirm" (Rui, 15).
d) Scanning several aggregators' news sites—"I surf multiple sites to confirm the information" (Beatriz, 14); "If it does not show elsewhere, it is not true" (Margarida, 12).
e) Seeing others' comments with regard to the news—"We see the comments of others. I always see the comments to see if it is true" (Pedro, 12).
f) Asking other people—ideally "parents" (Bruna, 12), but they also referred to their "brother" (Maria, 11), and "neighbors" and "cousins" (Duarte, 11).

Teachers reported that students were very informed by OSN but they considered that they do not confirm the sources, so the school should work on this area:

> I am fully aware that most kids do not do that. It is so on Facebook, and now with those false news sites there are many kids who read and think it is true. The school should have some caution regarding this issue and tell the kids they have to confirm the source. Tell them that the Internet is very good, of course, because it offers incredible possibilities, but we have to use it with some caution. (T7)

There is a contradiction between young peoples' perceptions regarding the confirmation of information and the perceptions that teachers have of the performance of these young people in this matter. Teachers advocate the need for the school to take action at this level, as they consider that there is a tendency for students to view the information they find as fact (Steeves, 2012a).

Teachers and Parents Use OSN as News Source

Teachers who consume information on the OSN stated that they always go to the primary source to confirm the information, whether on television or in newspapers, since "the information we find on Facebook is not enough" (T1), because "there is someone who does the screening for me; I do not like it" (T1), and "I do not trust Facebook news" (T25, T30, and T31). However, they give more credibility to the stories shared by friends in small groups: "If the news says that the data are from the National Statistical Institute and it is a colleague who shares in the group, you are not going to confirm from NSI" (T24).

Facebook operates as a portal: "I arrive home, I turn on my Facebook, and it is from there that I go to newspapers, to television, and make the management of news, since everything is connected" (T40) as "I have to just put 'like' in the pages of newspaper and television websites to receive the news" (T24). Those who claimed not to get information at the OSN preferred the news aggregators ("I do not watch news on the networks, I watch the SAPO"; T23), television, newspapers, and information services on mobile devices, considering that "Facebook is for fun, it is not for work" (T13). But they also got their information online.

Parents relied on the news "when there are other sources that confirm the information" (P20), so they often "confirm if what is being said on the network is truth" (P1) as "it has appeared, for example, on Facebook that a teacher evaluation test was cancelled when it was not true" (P11). And when the person or entity that shared information was a credible indicator: "I privilege information of people who I think are good sources of information" (P10).

Parents who did not get informed on OSN said they preferred "to go directly to news sites" because "sometimes the news that is on the networks is slow to open" (P2). They did not switch "from Facebook to websites" because "if you want to know some information, it is faster going to a news site" (P8). In any case, they also got informed online.

Publishing on OSN is an Adult Activity

Teachers were the group who published the most on OSN (77%), followed by parents (54%), ahead of young people (38%). Even among those who published on OSN, it was not a frequent activity, except for a third of the teachers, who stated that they published many days or every day, while young people and parents did not exceed 15% (Figure 5).

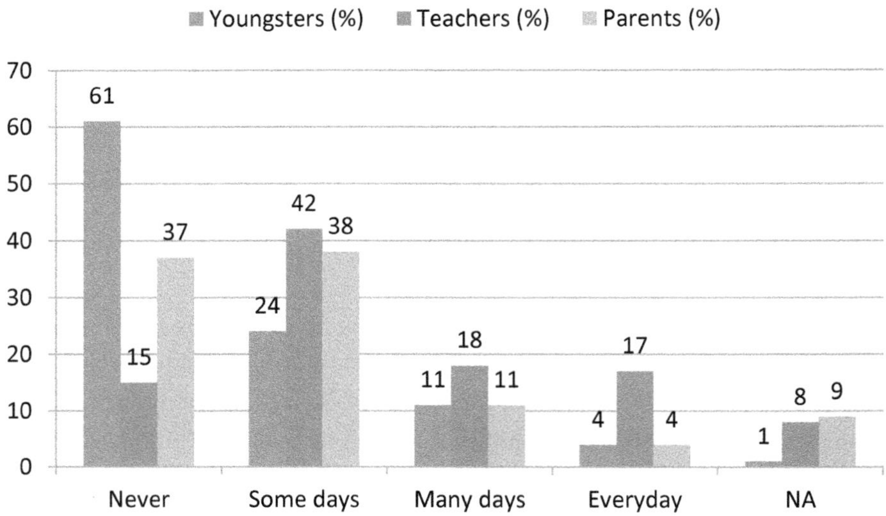

Figure 5. Publication on the network by young people, teachers, and parents (%) N = 499 (youngsters); 110 (teachers); 172 (parents).

Teachers claimed to publish for professional reasons, often in restricted groups, but most of their posts were personal, "especially image" (T10), such as "pictures of my son in football," since "when you post something that is meaningful to you, people put likes and comment" (T40).

The teachers interviewed considered the results to be exaggerated because they did not believe that 77% of their peers posted on OSN: "I think it is too much! I do not know if you noticed well the concept of 'posting.' Eventually they are mistaken with 'shares'" (T7), which justify the results being far above that of young people: "I believe that students know better than teachers the difference between "post" and "share." That is what I realize and because I share much more than I post" (T8).

A similar situation occurred with the parents, as "there may be some misunderstanding in terminology between what is 'posting' and what is 'sharing'" (P16), which was found. When we explained the difference between actions, it was common to hear phrases such as: "Oh, so I do not post. I share. I cannot even post" (P5). Parents who admitted posting on OSN emphasised "text, photos I take" (P3), "text and photos" (P20), "my personal tastes and opinions" (P18), or professional reasons ("I post on the company's page, disclose events"; P7). The rest posted "little" (P2) or did not publish.

The results for young people (38%) were similar to those found in Canada, where 41% declared posting on OSN (Steeves, 2014). In Europe, only 12% reported posting messages on websites or blogs and 17% admitted posting photos or videos (Mascheroni & Cuman, 2014), while in Portugal, they were respectively 9% and 11% (Simões et al., 2014).

The youngsters' posted particularly in the area of photography: "pictures of myself or friends, but only for friends" (Bruna, 12), "pictures of my friends taken in school" (Nuno, 12), "photos of my dog" (Joana, 12), "photos of myself and games" (João, 12), or "when I travel with my family, I put the coolest photos" (Maria, 11). From 13 years they stated that they posted more "in private groups... of friends, where we can talk about what we want" (Catarina, 15), but also published freely among participants with one or more common interests:

> Out of that group to which I belong, I do not post much. In the group, which is secret, I post more. We post almost everything: pictures, moods, our opinions. It is all centered on a band, but we can say what we want. (Elisa, 16)

Young people published "different photos on different networks" (David, 16) and sometimes added "a description, to say where it is" (Beatriz 14).

Concluding, publication on the OSN is not an adult action, unlike the results of the questionnaire. Teachers and parents may have difficulty distinguishing posting and sharing. Although in the questionnaire they were the least likely to post, here, young people were among the ones who posted the most, mainly photographs, with or without associated text. Publishing occurred mainly in private groups, from 13 years. But it is important to note that young people know the potential of technology and what they can do on OSN, which does not mean that they do so as, on the Internet, "only a very small proportion of users is, in fact, creating original content: most are simply 'consuming' content, as they always did" (Buckingham, 2012, p. 46).

Teachers, Youth, and Parents Share More Than Post

Teachers (75%) were the group who most declared that they shared information on the OSN, followed by parents (61%), and young people (40%). But it was young people who did it most often (29% did it many days or every day), although teachers (23%) and parents (21%) were close behind (Figure 6).

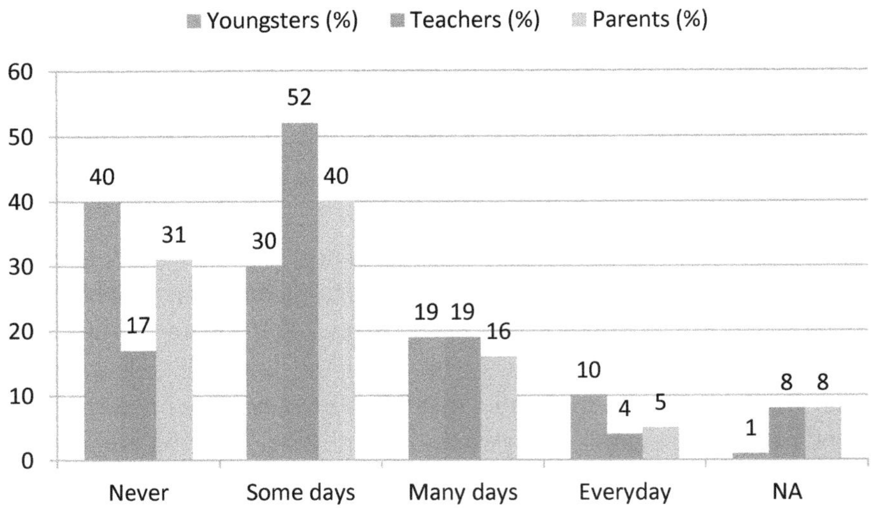

Graph 6. Shares on OSN by young people, teachers, and parents (%) N = 499 (youngsters); 110 (teachers); 172 (parents).

Teachers were those who most shared and did it selectively: "I do not share everything in the same place. I have several groups. When it comes to music, I go to a group. When it comes to concerts, I go to another. For family, I go to another" (T1). Examples of these shares are "paintings by famous painters and summaries of biographies" (T20), "music" and "stuff I find interesting" (T40), "news that has to do with our area [scientific]" (T24), or "colleagues' posts of the many photos of the places they go" (T3).

Sharing is also common for most parents, either "pictures of friends, some who I know only through Facebook" (P3), "news that is coming up" (P18), "articles related to work" (P14), "concrete matters, useful [issues]" (P17), or "jokes that have an associated learning" (P6).

Young people shared others' messages more frequently than their own, except for photos. The percentage of those who published someone else's videos (58%) more than doubles those who published their own (27%). They clearly shared more than they produced or published, despite the technological resources available. They shared music, photos, and outsourced videos in their profiles, not presenting them as someone else's (i.e., their own) property, so they did not have authorship concerns and, when faced with copyright issues, they alleged that any questions about source did not arise, and they even considered that the fact of sharing should be seen by the author as a compliment (Steeves, 2012b).

In the age groups 10–12 they shared "something that is funny" (Miguel, 11), "games for others [who are] starting to play" (Fabio, 12), "a lot of 'It is Ugly' [site of humor]" (Gonçalo, 11), "jokes" (Duarte, 12), football pictures or non-schoolwork they did: "Yesterday, I made three dolls and put them on Facebook, plasticized dolls" (Joana, 12). From 13 years they shared "video games" and "music" (Miguel, 13), "Football news" (Inês, 15), "anecdotes" (Miguel, 15), "extreme sports stuff...in those groups to which I belong" (Marcelo, 13) or "stupid videos, humorously mocking" (Rafael, 14).

Analyzing the data related to publishing and sharing, we identified three trends:

a) Publishing and sharing content are more common as we progress in age group, and all outsourced authorship of files: photo, video, and sound files. Still, at the age of 15, 21% did not publish photos of their own, 40% did not share photos found on the Internet, 35% did not share outsourced videos, and 40% did not share sound files of outsourced authorship.
b) There are more girls than boys publishing photographs of their own, as 70% of girls agreed that they did, against 60% of boys. And they did it with a higher frequency than boys (36% published many days or every day against 25% of boys).
c) The boys asserted that they posted more sound files than girls although with low percentages, as there are 31% of boys who admitted it, and 19% of girls.

To summarize, sharing is a more common action than publishing, among young people, teachers, and parents. The percentage of young people who stated that they share on OSN is lower than that of teachers and parents, perhaps because they distinguish better between publishing and sharing, while adults have more difficulty in doing so. Young people from older age groups tended to share in groups and not on their wall, as happens with their publishing. In these shares the image dominates, especially photography, just as with the publications, stressing that young people, with the girls leading, communicate more through image than text.

"Comment" is Synonymous With "Like"?

Teachers (74%) and young people (74%) commented more on OSN than parents (63%), but people do it many days or every day (39%), almost double the percentage of teachers and parents (Figure 7).

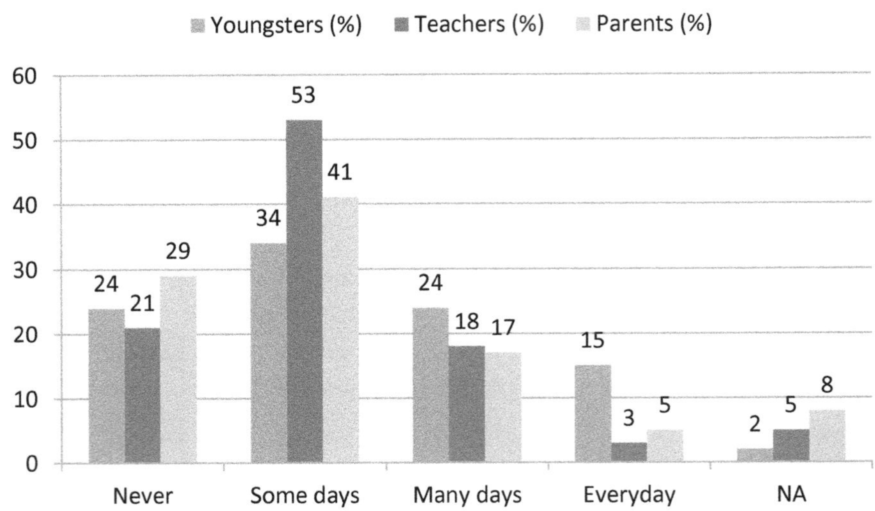

Figure 7. Comments' frequency in OSN by young people, teachers, and parents (%) N = 499 (youth); 110 (teachers); 172 (parents).

Teachers were organized into two groups: those who commented openly and those who did so with caution and sometimes privately.

For the first group, the comments were necessary "so that people know that they are seen by others, as it is irrelevant being in a network if there is no communication" (T1). They did not fear exposure on the OSN, but simply marked "like" ("I mark many likes. If I am on the network and see that it is marked like, I do it"; T19), and they believed that most colleagues did the same, in some cases with associated comments: "When I have time, I mark 'like' and comment. And my colleagues do it too" (T8).

Among the second group, precautions were common ("If I comment it is via private message and only if it is with my daughter"; T25), invoking self protection and the protection of others: "when people post moods...I do not comment directly, but I [send a] message, as other people do, especially when there is closeness" (T40).

Commenting was rare, as "commenting is only sporadically and with people I know personally" (T13), "people I trust" (T24), or avoiding the hackneyed "like": "Sometimes I mark 'like' here or there, but scarcely" (T14) since "I do not like to always put 'like' because everyone can see and tell that I was on the network" (T24). The few parents who stated that they comment did it frequently ("I easily react to what they post on Facebook"; T5), citing as examples "news of public interest" or "those [things] more mundane, posted by friends" (T14), especially if the "like" is considered a comment:

> Most people do this single action. Go through content which pleases them and mark "like" in the sense of saying you were there and you saw. If we consider that mark as a comment, it does not surprise me at all that parents assume themselves [to be] interventional or active. (P9)

The "like" is also the most common action among young people ("I just do like or dislike. No more comment[s]"; Miguel, 11), sometimes associated with telegraphic comments: "I do 'like' and write in comments such as 'Big dung'" (Gonçalo, 11), or "[for] a funny thing, we put LOL" (Joana, 12).

In the age groups 13–15, the "like" dominates the preferences and is a bargaining chip, but with less value than the comment:

> For example, a girl posts a photo on Facebook. Only if she is cute, we comment and say "You are so beautiful" and such....If not, we do not care. We mark a "like" because then we also get one. (David, 15)

The "like" is something directed to contacts in general, while the written comment means/implies more closeness:

> People only comment when they have more affinity, when they really know each other, I will not comment on a thing of a person who I do not know. I put "like," so like saying, "I saw your post and liked it." (Liliana, 16)

Comments are more common and made in an uninhibited manner in closed groups: "Those groups to which I belong, if I know some of the administrators I usually comment enough" (Marcelo, 13).

Young people commented on OSN when their name came up associated with a post (77%), when identified in a photo or video (77%), but also when referring to friends who know them personally (68%) or when they want to give an opinion on a subject that interests them (72%). Messages posted by their idols (e.g., football players or TV stars) motivated the comments of around half of young people (55%) and only 21% commented on the posts of the profile contacts who they only knew virtually. Although three quarters did not comment on posts from people who they did not know personally, boys are the group who did it more (28% of boys but only 16% of girls).

Some types of publications were more commented on by young people as they progress in age group, with significant differences, as young people of 13–15 were those who commented most when "my name is in the message," "I am identified, in [it]," "when the subject interests me and I want to give my opinion," and "when it concerns a friend who I personally know." There were also young people who declared that they commented more in closed groups, which helps to explain the increased frequency of the comment.

The OSN as the Preferred Way to Communicate

Young people are the group who most claimed that they talked to friends through OSN (96%), followed by teachers (72%), and parents (46%) (Figure 8).

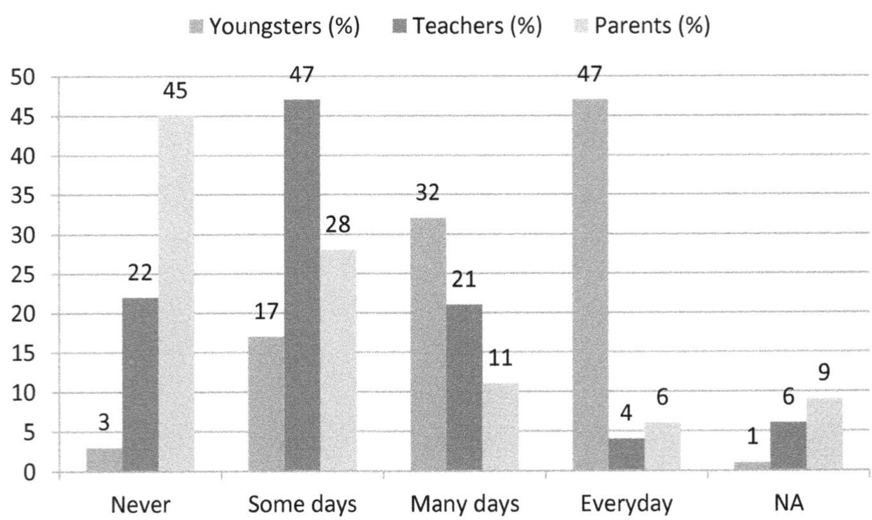

Figure 8. Frequency with which young people, teachers, and parents talk to friends on the OSN (%), N = 499 (youngsters); 110 (teachers); 172 (parents).

Three out of four youngsters talked with "friends" (we use the Facebook terminology here) on OSN many days or every day, a frequency only mentioned by a quarter of teachers and 17% of parents. Half of the teachers and a quarter of parents declared that they spoke with "friends" through the OSN sporadically (some days).

The teachers interviewed were divided into three similarly sized groups: those who used the OSN to speak with "friends," those who spoke but preferred to use other media, and those who did not speak through OSN:

a) OSN users: some did regularly and for work reasons ("I talk to many colleagues, even on school subjects at night....It has happened to me [that we] get out of school and half an hour later we are talking on the network"; T7), as was the case of a music education teacher who used OSN to "schedule tests among musicians" (T1). But sporadic contact was the most common: "There are people to whom I say 'Hi' every day, but I see them often. Talk is more a weekly thing" (T40);
b) Less frequent OSN users: conversations through the OSN occurred "very occasionally" (T13), "from time to time, more on holiday" (T14), as "colleagues communicate more by phone and even e-mail, which is more private" (T8);
c) OSN non-users: they blocked chat applications "because I never bothered to be available only to A, B, or C and was paranoid. I had no ability to talk to so many people and I turned it off" (T10), or not to disturb their work: "Now I have the chat always disabled. Formerly I ended classes at midnight and went to bed at two in the morning. And it appeared to people who I barely knew, to meddle in conversation" (T24).

Parents can also be organized into three groups:

a) OSN users for convenience: "I use it to talk to my son who is abroad" (P4); "There are people who we do not get even standing in front of them. They do not listen to us. If we get to them in a more natural way [for them], it is more effective" (P41);
b) Less frequent OSN users: "I can find someone there and if I have something to say, I can take advantage" (P7), "I hate the chat, I do not use it, except for a very important thing but it is sporadic" (P2);
c) OSN non-users: they preferred "talking personally" (P19), or because having the chat open bothered them, as mentioned by the

> teachers: "Most of the time I turn off the chat to avoid it bothering me, or I do not do anything else other than talk" (P17).

Young people were permanently connected. The OSN are the ideal platforms, complemented by SMS. The permanent connection does not replace face-to-face socialization, "but instead of stopping at the doorway of their homes, it continues through the constant texting which can be done at all hours" (De Abreu, 2011, pp. 47–48):

> Y8—First we have to say something that we have forgotten or something like that. Then it begins a conversation. Most often [they] are stupid conversations, where no one is treating any issue. Nonsense, whatever!
> INT—But is that permanent?
> Y3—Sometimes we fell asleep and then wake up, for example, at four in the morning, and there I have a message. I answer.
> (Laughs of approval)
> Y5—Ya, then the conversation goes on.
> Y1—Sometimes we have conversations that last for days.

The nonstop use and even overnight use was confirmed by the teachers: "There are children who come to school tired because they were chatting until two in the morning....And parents think they are asleep" (T1). The statements of the young and the perceptions of teachers point in the same direction as a British study that surveyed 848 young people 12–15 about their sleep habits and found that about a quarter were connected to the OSN permanently and agreed that they often wake up during the night to use them, so they feel tired when they arrive at school in the morning (Horton, Taylor, & Power, 2015).

Regarding the mobile phone, it enables users to communicate via SMS even when they do not have access to the Internet, thus ensuring the permanent connectivity ("For example, the lovers that are always talking by SMS, even if they have Facebook"; Maria, 15) with the advantage that "almost everyone does not have to pay" (Rafael, 14). It is the way used to establish conversations that would be more difficult to have face-to-face or via video call:

> Y4—Francisco in our class, his girlfriend broke up with him by SMS.
> Y5—With my last boyfriend it also happened [like] that.
> Y4—She should be afraid that he started crying.

Y3—No doubt there is some bad news, or so, that I do not have the courage to say face-to-face. I do not like talking to people by SMS when it happens. But it is easier to say a bad thing that happened.
Y7—Yes, it is easier. You feel freer, is [that] not it?
Y8—We feel safer and if the person does not want to remember, just delete the message below. So we do not take a slap next and so by SMS is easier.

The speed of communication is crucial, and here the OSN are important because "in the chat you can check if the person saw [the message], or not. It is faster" (Maria, 16). They also used Skype to talk "with family members who are far away" (Leonor, 14). Skype and OSN are crucial to "agree [on] gaming strategies" (Marco, 11) while playing, avoiding using the in-game chat: "When we are playing, we talk a lot in the network chat. Thus others do not realize that we are talking among us" (Miguel, 11). The use of e-mail was rare ("I use the e-mail more with family to send photos or so, but with my cousins [it] is already by chat"; Rui, 16), mainly to send schoolwork.

Communicating Through the OSN

Young people use technologies such as OSN "to extend the relationships with those who are already part of the different circles [in which] they participate" (Endrizzi, 2012, p. 7) and to maintain multiple and constant lines of communication with their closest friends (Ito et al., 2009).

Almost all (97%) talked to friends who they knew personally through the OSN. As they progressed in age group, they declared doing so more. If at age 10, 11% of young people did not speak with "friends" on the OSN, from 11 years, the percentage of those who did not do so is not beyond 5%. The frequency of conversations also increased with age, since those who talk many days or every day are 53% at 10 years, 78% at 11 years, and about 90% from 13 years forward.

Among the "friends" who they personally know are:

a) Schoolmates (mentioned by 86%)—"We talk more with school friends and not so much with classmates" (Juliana, 13).
b) Family members—Over 70% claim to speak "with relatives or friends living abroad."
c) Boyfriends or girlfriends—These conversations are undertaken by 39% of young people, with 47% of boys, and 32% of girls claiming to do it. But girls disagree with the results: "I think many people

> included boyfriends in friends. They do not say boyfriends, perhaps because they do not even consider them boyfriends. But it is as if they were" (Rita, 16). The frequency of conversations increased with age, as the percentage of those who do it on a daily basis grows from 8% to 29% between 10 and 14, reaching 43 % at 15.

These contacts were even more common among young people who accessed via mobile devices, and within these, those who had their own for private and personal use, such as a smartphone (Mascheroni & Cuman, 2014; Simões et al., 2014; Steeves, 2014). This was true among the young people surveyed. Those who had mobile access spoke more with school friends even after they came home than the young people who accessed the Internet via personal computer or family PC (89% compared with 85%). And 37% of youngsters having mobile access stated they do it every day against 27% of those who did not have mobile access.

A third (32%) of young people talked with friends who they did not know personally, and of those, it was more boys (40%) than girls (27%). There were also young people from older age groups who did it more frequently. At 12 years it was less than 30%, but the percentage increased to 38% at 13, 46% at 14, and 58% at 15. "There are many people who meet on Facebook. So I think it makes sense" (Mariana, 15). Since the number of new contacts on the OSN increased with the age of the young people, it was understood that the conversation increased with people who they did not know personally, which can be perceived as new contacts, thus fulfilling the principle that on the OSN and in terms of contacts, "the rich get richer," for those who had more contacts got more new contacts and got them faster (Recuero, 2009, p. 67).

Main Conversation Issues Among Young People

School subjects (69%) and conversations about different subjects ("everything and nothing"), which 60% elected, were the most-mentioned topics of conversation held on OSN. All other issues were reported by about 45% of young people, who talked about other media and their contents through the OSN, such as cinema, music, television, or video games, informatics, and sports. The OSN was still used to date and to talk about dating (41%).

Boys spoke more about sports, video games, and computing. Girls spoke more about school subjects, about "everything and nothing," culture, dating, and current topics. Activities such as making appointments with friends, talking about "everything and nothing," and talking about dating or current topics were referred to more as we progressed in the age group.

About 97% said they used the Portuguese language in those conversations, followed by English (40%), while 10% (50 youth) use other languages, such as Spanish (18) and French (15), followed by German (four), Italian (three), Ukrainian (two), Russian (two), Mandarin, Korean, Japanese, Dutch, Sindarin, and Kenyan. Of the total OSN users, more than half (55%) used only Portuguese, while 42% used Portuguese and other languages.

Although there was no difference between genders, young people from older age groups used other languages on OSN, with significant differences among age groups. If up to 13 years, the use of other languages was about 7%, at 14 years 10%, and at 15 years 17%. Girls were most likely to use the English language, as 47% said they did it, against only 33% of boys.

Even young people who did not use the English language on OSN referred to its importance: "English is something essential; as it is a universal language... everyone can understand. But I only post in Portuguese because I do not know anything about English" (António, 15). Those who used the English language presented six reasons for doing so:

a) It is a universal language "everybody can understand" (David, 15) like their "English friends" (Bruna, 13) or "virtual friends, who are not all Portuguese" (Inês, 13). English is used more on other OSN than Facebook: "On Twitter and Tumblr we write more in English. Facebook is more our friends, is more Portuguese....In the other networks there are people we do not know, people of other nationalities" (Vera, 14).
b) It enables them to reach bigger audiences: "When [it is] international stuff like most of these things are in English, we comment in English so other people can understand" (Nuno, 12).
c) It is used "because it gives more likes" (Paulo, 14), "even when mixed up with" the Portuguese language (Marina, 13), since "the English language is more beautiful and it is something else" (Helena, 14).
d) It is the most appropriate language to express feelings, that is, it "is much more quaint and more sentimental" (Maria, 13), since "the words seem to gain more sense" (Reinaldo, 13).
e) It is the language of the videos because "in fact, who is going to watch videos in Portuguese? Anyone. Just in English. And there are things in English that do not make sense in Portuguese" (Marta, 13).

f) It is the language of online gaming, in which they have teammates or opponents with other nationalities, so they develop communication strategies: "When I play, I go to Google [Translate]. I put the word out there and the English word appears" (Gonçalo, 11). The translation is not always noticeable, so they appeal to others; "Google [Translate] already deceived me sometimes. But I have an English friend who speaks Portuguese. I send him a message with the sentence in English and he translates it [for] me" (David, 11).

The OSN associate the local context to the global one, so users "express themselves through a mix of the mother tongue and international languages, particularly English" (Enli & Thumin, 2012, p. 95). The relationships are mediated in a different logic from local or national media, as communities in which people integrate are based on social relations and not on geographical territories.

Three Participation Profiles

The individuals of the three sub-samples had the same interest in getting informed on OSN and differed in communication through the OSN, as all young people preferred them, while adults were distributed into those who did, those who admitted doing it sporadically, and those who did not, whether by choice or not.

Regarding the activities of publishing, "sharing," and "commenting," we defined three profiles we consider to be common to the three sub-samples. We relied on Enli & Thumin (2012) who, as a result of a longitudinal study to analyze the participation of Norwegian university students in OSN, considered three profiles: (1) the "reluctance practice," which shows a minimum of information but never anything private, intimate, or sensitive; (2) the "promotional practice," which corresponds to a transition phase, in which the individual publishes positive information about oneself, a kind of public relations, which guarantees some feedback; and (3) the "sharing practice," which involves sharing innermost, emotional, and private information, albeit with limitations, but which provides power and guarantees that others will provide feedback.

Thus we identified three profiles: Voyeurs, Beginners, and Interactives:

a) **Voyeurs**—People who did not create/maintain the profile to interact with others, but only to give that feeling. They were in the OSN especially to see what happens, so they did not comment, did not publish, and rarely shared. "I do not share or publish... I only see what others put there, what others comment and that" (Catarina, 12); "I do not use Facebook for this [publish, share, or comment]. It is just to chat and see what happens" (Márcia, 15).
b) **Beginners**—People who used OSN with more fear than confidence, so they did not publish, rarely shared, and the comment received was nearly always "like," unless it was done by private message. They got more than they gave. "I do not publish every week or every month. Most people publish news that may have to do with our area, or funny thoughts, and publish many photos of the places they visit, but I do not....And I do not like to put always 'like'" (T3).
c) **Interactives**—People who communicated, shared, or commented more, ended up developing the three actions, could distinguish from one another the comment, "like," post, share, and who made different comments on pages, groups, or walls, according to the recipients. "I post every day because I have a blog associated with Facebook. And I share a lot, not ordinary things or trash, but specific and useful things. I comment especially on news or posts of friends" (P14).

As participation in OSN tends to be dynamic and evolving among young people, we admit that evolution can also occur among adult users of OSN, who can develop, or not, from the first to the third profile. We use, therefore, the profiles outlined by Ito et al. (2009) to characterize the forms of online participation that are first organized into two groups. One is based on the common practices of relationships with friends and peers ("friendship-driven") and the other is the development of practices toward specific interests or hobbies ("interest-driven"). Depending on the involvement of individuals and frequency of practice, researchers established three subgroups: (1) the first develops activities such as gaming, chatting online, and participation in OSN to make friendships ("hanging out"); (2) the second, which corresponds to a transition phase, is the beginning of a more intense involvement, doing research and playing with the media, posting some videos and photographs ("messing around"); (3) the third involves a specialized participation, a frequent and intensive use, and a desire to produce media and intervene socially through the media ("geeking out"). Due to the progress we perceived between the questionnaire survey and the interviews, the evolution is not unique to children and youth, but also includes adults.

The Youth's "Friends"

Approximately 60% of young people surveyed had up to 300 contacts on OSN, 16% up to 500, and 24% had more than 500. The number of contacts increased with age.

Digital "friends" tend to match with those they know physically, because 41% of young people stated that they knew all their contacts on OSN personally, 42% said they know most, only 8% knew half, and 10% less than half.

The greater the number of "friends" on OSN, the lower the number of those they personally knew, which is valid for boys but even more so for girls. The girls, especially at 14 and 15, tended to have more contacts on OSN as a profile with many contacts "means popularity and implies a degree of social capital, particularly among young people and celebrities" (Enli & Thumin, 2012, p. 93).

Boys admitted that girls have more "friends," especially "if they are beautiful" (Manuel, 12) and/or if "they post photos that call boys' attention" (Rui, 15), since "boys add girls by the profile photo; if it appears [to be] a charming girl, I ask for friendship and the others reciprocate" (David, 15). There are girls who "add everyone" (Joana, 12) because "they want to have as many ['friends"] as possible" (Fábio, 12), since "they are always checking who has more 'friends' and telling the others" (Gonçalo, 11). From the boys' perspective, "girls are more troublemaker[s] and want to be always better than others" (Carlos, 11). From the age of 13, "the competition is more on the number of followers than the number of friends." "The number of followers on Tumblr and [ASKfm] is more important than the number of friends on Facebook" (Inês, 13). The same happens on Facebook: "Now the number of friends is less important. As there are followers, we may have few friends but have many followers" (Maria, 15).

Although half of young people admitted having contacts on OSN who they did not know personally, if we look at the reasons why they accepted them, we find out that they were not complete strangers: they were attending the same school (31%), friends of friends (17%), or lived nearby (8%). The rest were indeed strangers, and we accepted them because they were girls/boys (17%), they seem friendly in the photo (11%), they had the same interests (6%), they were the same age (5%), or partners in online games.

Among the "friends" of three quarters of youngsters surveyed are parents and/or their teachers. A third of young people claimed to be a friend of parents and teachers, but among those who had "friends" who were parents or teachers, the parents were the most common (31%) (Figure 9).

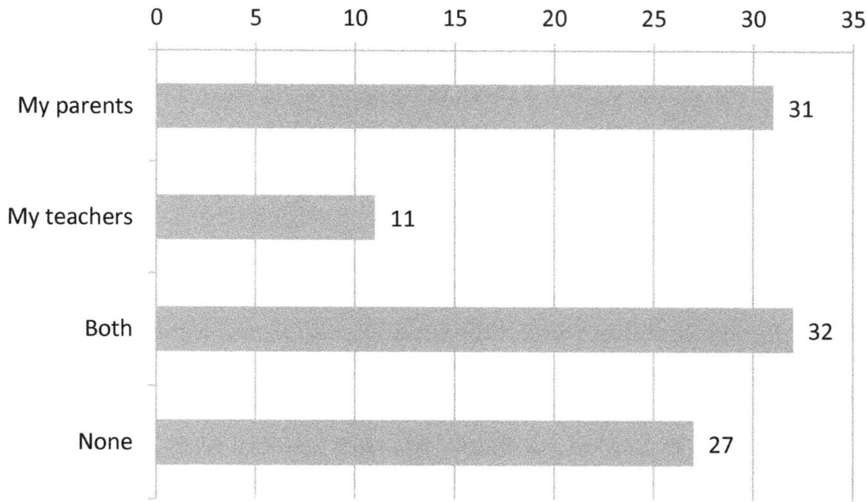

Figure 9. Young people that are, or are not, friends of parents and/or teachers (%) N = 499.

A quarter of young people had no teachers or parents among their contacts on OSN because their parents had no profiles on OSN (66%), they did not want parents or teachers among their "friends" on OSN (39%), or because teachers did not accept their request (3%). Other reasons (8%) were "my parents are not in my network," "as a matter of privacy," and "we should not send requests of friendship to teachers." Young people claimed that parents were "inquisitive" (Manuela, 16) and "cocky" (Pedro, 14).

Young people from older age groups had more teachers (and not parents) among their profile contacts (3% at 10 years and 18% at 15 years). They were mostly former teachers because "[when] they are our teachers [they] do not accept us, but when they are no longer, [they] accept" (Vera, 14). Teachers had more former students than current ones among their profile contacts (Morduchowicz, 2014), because the access of young people to the teacher's profile reduces the boundary between the person with authority and the private/personal self, which requires reflection by the teachers about accepting students or not in their profile contacts (Süss & Hipeli, 2013). If only a small group interacted with the students from "a professional profile, as they do not know that I have a personal one" (T23), most did it from a personal profile: "I would like to have a professional profile for that, but I have no time so I use my private one" (T5).

Young boys are the ones that most stated that they included parents in their profile contacts (35% against 26% of girls). One reason was online games: "Sometimes parents have Facebook because it helps us in games,… [when] you need to send some things to others, to 'have energy' and things like that. I created Facebook for them because of it" (Pedro, 11).

"Being a Friend" Does Not Mean "Talking To"

Four in 10 young people (42%, 207 individuals) talk with parents on OSN, especially the group 10–12 years. But in the focus interviews, only a few admitted doing it and those who did pointed to casual and infrequent talks: "I sometimes talk to my mother, if she is online" (Marta, 12); "I sometimes talk to my father, if I have no one else to talk [to]" (António, 12); "Once I was sick and asked my father, on Facebook, whether I should go to school or not" (Elisa, 16). The lack of contact is justified: "Parents are with us at home, so we do not need to talk to them" (Carlos, 17). The results are consistent with those obtained in the Portuguese part of a European study, as 77% of young people surveyed said they never or almost never contacted parents through their profile on OSN (Ponte & Simões, 2015).

Of the 172 parents of OSN users, 76% claimed to be their children's "friends," 11% reported that they were not, and 13% did not answer the question. But being a "friend" does not mean that they "talk to" [them online] because only a third of parents that were their children's "friends" talked to them sporadically through the OSN. Parents claimed to "[casually] talk" because "once in a while she goes to the school library, gets on Facebook, and says 'Hello'" (P18), "if [we] needed [to] call her for dinner on Facebook (laughs)" (P41), or by reasons economic and of convenience: "When he went to Spain with my sister, we spoke on Facetime" (P4). If parents needed to talk to students they preferred the mobile phone, "because she reacts more to SMS" (P16).

Only 21% (108 individuals) of young people claimed to talk sporadically to teachers through the OSN: "The Science teacher last year congratulated me on my birthday. I also sent her a message, but we never talked on the chat" (Alexandra, 13). Only 5% talked many days or every day. Conversations with teachers were mainly an exchange of private messages and not effective dialogues on chat, with the recurrent phrase, "I do not talk to teachers."

Of the 110 teachers who use OSN, 67% had students among their profile contacts. But only 40 (36%) talked with students and they did it sporadically. The talks had pedagogical objectives (31 cases), were of support/supportive—i.e., were a good use of the OSN—(20 cases), and were used to deal with personal matters (20 cases). "Other" objectives were "enlightenment and activities advertising," "cultural," or "talking about matters of day-to-day at school." Teachers did not take the initiative: "If students say something, I answer" (T5). Those who took the initiative established single-direction contacts: "if it is someone's birthday, I congratulate them. Nothing else" (T40), and some contacts were "for fun, informal talk, casual" (T1). Even conversations related to school content assumed these features:

> Look, some of them send me messages like, "Oh teacher, when do you post the exam matrix?" I do not usually talk, but I talk. Some know that I am there and it is not the first time they [have] ask[ed] me things on the subject, doubts and that kind of stuff. It is not just clear doubts but sometimes [they] ask whether "some part of the subject is included" or "a book page." All simple questions that you answer quickly. (T24)

Chats between teachers and parents through the OSN were even rarer. Only 9% of parents (15 individuals) said they had spoken to their students' teachers, particularly about school issues related to their student (10 cases), the use of the OSN at school (four cases), personal issues (four cases), and the use of the OSN by the student (one case). Only one of the 20 parents interviewed admitted talking to his children's teachers through OSN because he knew them personally, and because it was about school issues.

Conclusions

Nine out of 10 young people, three out of four teachers, and two out of three parents used OSN and Facebook was the most used in each of the three groups, followed by YouTube and Google+. Twitter was mentioned most by youngsters and Flickr assumed residual values. About 40% of young people and a third of teachers and parents used Facebook every day, although they had a profile on others (e.g., Tumblr, ASKfm, Instagram) that perhaps they used more. On average, young people and teachers had profiles in three OSN and parents in two. Among young people, the use increased with age: at 10 years already three in four had a profile on an OSN, but at 14 and 15 years they were more than 95%. About 90% of young people started using OSN before 13 years and those who did not use it were justified by the lack of access to the Internet or prohibition from parents.

The OSN are information spaces par excellence as teachers (89%), parents (84%), and young people (77%) declared that they became informed on OSN, and about half did it many days or every day. The percentage of young people who got informed on OSN increased with age and, at 14–15 years, over 80% said they got informed on OSN, and over 60% did so several days or every day.

The OSN, integrated into the logic of Web 2.0, more than the consumption of information, enables its creation and publication (Rivoltella, 2012), but this potential is under-exploited (Buckingham, 2012). Three out of four teachers and half of parents claimed to publish sporadically on OSN, but in interviews they made clear the confusion between publishing their own productions and sharing third-party productions. Participation in OSN means more commonly sharing third-party content (stated by about 80% of teachers, 70% of parents, and 60% of young people), rather than sharing their own content. Regarding young people, participation in the OSN was based on the image, as they published mainly photographs (63%), shared their own videos (27%), shard photos (67%), and shared videos of third-party authors (58%), and the girls did it more and more often as the age increased.

Three quarters of teachers, parents, and young people declared that they commented on posts on OSN, but the interviews indicated that these comments were mainly Facebook "likes" or its equivalent, which works to point out the presence among adults, but which means popularity among young people: the more "likes" their photographs have, the more popular they become, so these comments are important in their social validation process (Livingstone, 2014) by their OSN contacts. Despite their wish for popularity, which among the eldest is measured more by followers than the number of contacts, as age increases, so does the tendency of young people to publish, share, and comment in closed groups that adults and other peers do not have access to.

In terms of publishing, sharing, and commenting on OSN, we have identified three profiles: individuals who are on OSN to maintain, establish, or strengthen bonds of friendship and to observe what is happening (Voyeurs); those who have some involvement with content and public interaction, although weak (Beginners); and those who participate effectively because they know how to do it and have specific interests (Interactives). The passage through the three phases represents an evolution of the relationship of young people with OSN (Ito et al., 2009), but data indicate that the same facility exists among adults.

OSN are communication areas for young people (96%) and teachers (72%), but less for parents (46%). Young people seek to be always connected, whether day or night, whether in the classroom, at home, or on the street. They want to be in touch with friends, family, and boyfriends or girlfriends. Adults are divided among those who actually use the OSN to talk (especially for work reasons or to contact distant relatives), those who prefer other means such as the phone or face-to-face contact, and those who do not do so because they have used the OSN in the past, and felt that staying in touch can be a nuisance and a waste of time.

About two thirds of young people had parents among their contacts on OSN and 43% teachers; 76% of parents had students among their contacts while 67% of teachers had students in their contact lists. But being among the contacts does not mean interacting, since they can practically ignore each other. Although 42% of young people and a third of parents claimed to talk to each other on OSN, they did so casually with a simple "Hello," except in special circumstances such as travelling or when on vacation. And if 36% of teachers claimed to speak with students, few did it with educational objectives such as the clarification of doubts. Avoiding contact by teachers in relation to students could be in relation to the fact that they are connected through personal profiles and not professional ones, so the teachers are exposed. Thus it is understandable that they related primarily to former students, which is what students confirmed. Contacts between teachers and parents through the OSN were rare and only when they knew each other personally, because it usually happened via e-mail.

References

boyd, d. (2015). *É complicado—as vidas sociais dos adolescentes em rede*. Lisboa, Portugal: Relógio d'Água.

boyd, d., & Ellison, N. (2007, October). Social network sites: Definition, history, and scholarship. *Journal of Computer-Mediated Communication*, *13*(1), 210–230.

boyd, d., Hargittai, E., Schultz, J., & Palfrey, J. (2011). Why parents help their children lie to Facebook about age: Unintended consequences of the "Children's Online Privacy Protection Act." *First Monday, 16*(11). Retrieved from http://firstmonday.org/ojs/index.php/fm/article/view/3850/3075

Buckingham, D. (2012). Precisamos realmente da Educação para os Meios? *Comunicação & Educação, 17*(2), 41–60.

Cavazza, F. (2015, June 3). Social media landscape 2015. Retrieved from http://www.fredcavazza.net/2015/06/03/social-media-landscape-2015/

Chaudron, S. (2015). *Young children & digital technology: A qualitative exploratory study across seven countries*. Luxembourg City, Luxembourg: Publications Office of the European Union. Retrieved from https://ec.europa.eu/jrc/en/publication/young-children-0-8-and-digital-technology-qualitative-exploratory-study-across-seven-countries

CLEMI. (2013). *Médias & Information: On apprend (édition 2013–2014)*. Paris, France: Author. Retrieved from http://www.clemi.org/fichier/plug_download/54153/download_fichier_fr_medias_et_information2014_15.pdf

ComScore. (2015). *US digital future in focus 2015*. Retrieved from https://www.comscore.com/Insights/Presentations-and-Whitepapers/2015/2015-US-Digital-Future-in-Focus

Creswell, J. (2014). *Research design: Qualitative, quantitative and mixed methods approaches*. Thousand Oaks, CA: Sage.

De Abreu, B. (2011). *Media literacy, social networking and the Web 2.0 environment for the K–12 educator*. New York, NY: Peter Lang.

Endrizzi, L. (2012). Jeunesse 2.0: Les pratiques rélationelles au coeur des médias sociaux, *Dossier d'actualité Veille et Analyse, 71*, 1–24.

Enli, G., & Thumin, N. (2012). Socializing and self-representation online: Exploring Facebook. *Observatório*, 6(1), 87–105.

Fletcher, R., Radcliffe, D., Levy, A., Nielsen, R., & Newman, N. (2015). *Reuters Institute digital news report 2015 supplementary report*. Oxford, England: Reuters Institute for the Study of Journalism.

Gee, J. (2010). *New digital media and learning as an emerging area and "worked examples" as one way forward*. New York, NY: MIT Press.

Hobbs, R. (1998). Literacy in the information age. In J. Flood, D. Lapp, & S. Brice Heath (Eds.), *Handbook of research on teaching literacy through the communicative and visual arts* (pp. 7–14). New York, NY: International Reading Association, Macmillan.

Horton, K., Taylor, C., & Power, S. (2015, September 16). *Routines and rest: The sleep behaviours of 12 to 15 year olds*. Paper presented at the British Educational Research Association Annual Conference, Belfast. Retrieved from http://www.wiserd.ac.uk/news/latest-news/want-get-good-nights-sleep-kids-turn-facebook-and-twitter-say-researchers/

Individuals using the internet for participating in social networks—% of individuals aged 16 to 74. (2015, September 28). Retrieved from http://ec.europa.eu/eurostat/tgm/table.do?tab=table&init=1&language=en&pcode=tin00127&plugin=1

Ito, M., Horst, H., Bittanti, M., boyd, d., Herr-Stephenson, B., Lange,…Robinson, L. (2009). *Living and learning with new media: Summary of findings from the Digital Youth Project*. New York, NY: MIT Press.

Jenkins, H. (2006). *Convergence culture: Where old and new media collide.* New York, NY: New York University Press.

Jenkins, H. (2009). *Confronting the challenges of participatory culture: Media education for the 21st century.* New York, NY: MIT Press.

Kaplan, A., & Haenlein, M. (2010). Users of the world, unite! The challenges and opportunities of Social Media. *Business Horizons*, 53(1), 59–68.

LeDeuff, O. (2011), Éducation et réseaux socionumériques: Des environnements qui nécessitent une formation, *Hermès*, 59, 67–73. Retrieved from http://www.cairn.info/article.php?ID_ARTICLE=HERM_059_0067#no12

Lovink, G. (2011). *Networks without a cause: A critique of social media.* Cambridge, England: Polity Press.

Lopes, P. (2015). "Muito entretenimento, pouca informação", European Journalism Observatory, Retrieved from: http://pt.ejo.ch/10108/literacia-dos-media/nativos-digitais-muito-entretenimento-pouca-informacao.

Livingstone, S. (2014), Developing social media literacy: how children learn to interpret risky opportunities on social network sites, Communications, 39 (3), 283–303.

Mascheroni, G., & Cuman, A. (2014). *Net children go mobile: Final report.* Deliverables D6.4 & D5.2, Milano, Italy: Educatt.

Mihailidis, P. (Ed.). (2012). News literacy: Global perspectives for the newsroom and the classroom. Peter Lang.

Morduchowicz, R. (2014). *Los chicos y las pantallas: Las repuestas que todos buscamos.* Buenos Aires, Argentina: Fondo de Cultura Económica de Argentina, SA.

Obercom. (2015). *A Internet e o consumo de notícias online em Portugal.* Lisboa, Portugal: Observatório da Comunicação—Obercom.

Ólafsson, K., Livingstone, S., & Haddon, L. (2014). *Children's use of online technologies in Europe. A review of the European evidence base.* London, England: EU Kids Online.

O'Neill, B., & Staksrud, E. (2014). *Final recommendations for policy.* London, England: EU Kids Online, LSE. Retrieved from http://www.lse.ac.uk/media@lse/research/EUKidsOnline/EU%20Kids%20III/Reports/D64Policy.pdf

Panek, E., Nardis, Y., & Konrath, S. (2013). Defining social networking sites and measuring their use: How narcissists differ in their use of Facebook and Twitter. *Computers in Human Behavior*, 29, 2004–2012.

Pew Research Center. (2015). *Teen, social media and technology overview 2015.* Washington, DC: Author.

Ponte, C., & Simões, J. (2015). Comportamentos *Online* de Crianças e Jovens Portugueses. In G. Miranda (Ed.), *Psicologia dos comportamentos online.* Lisboa, Portugal: Relógio D'Água Editores.

Recuero, R. (2009). *Redes sociais na Internet.* Porto Alegre, Portugal: Sulina.

Rivoltella, P. (2007). Realidad y desafíos de la educación en medios en Italia. *Comunicar*, 28, 17–24.

Rivoltella, P. (2012). Retrospectivas e tendências da pesquisa em mídia-educação no contexto internacional. In M. Fantin & P. Cesare Rivoltella (Eds.), *Cultura digital e escola: Pesquisa e formação de professores.* Campinas, Portugal: Papirus Editora.

Rosado, A., & Tomé, V. (2015). As redes sociais na internet e suas apropriações por jovens brasileiros e portugueses em idade escolar. *Revista Brasileira de Estudos Pedagógicos*, 92(242), 11–25.

Simões, J., Ponte, C., Ferreira, E., Doretto, J., & Azevedo. C. (2014). *Crianças e meios digitais móveis* em Portugal: Resultados *nacionais do projeto net children go mobile*. Lisboa, Portugal: CESNOVA.

Statement of rights and responsibilities. (2015). Retrieved from https://www.facebook.com/legal/terms

Statista. (2015). *Leading social networks worldwide as of August 2015, ranked by number of active users (in millions)*. Retrieved from http://www.statista.com/statistics/272014/global-social-networks-ranked-by-number-of-users//

Steeves, V. (2012a). *Young Canadians in a wired world, phase III: Teachers' perspectives*. Ottawa, Canada: Media Smarts. Retrieved from http://mediasmarts.ca/ycww/teachers-perspectives

Steeves, V. (2012b). *Young Canadians in a wired world, phase III: Talking to youth and parents about life online*. Retrieved from http://mediasmarts.ca/ycww/talking-youth-parents-about-life-online

Steeves, V. (2014). *Young Canadians in a wired world, phase III: Life online*. Ottawa, Canada: Media Smarts. Retrieved from http://mediasmarts.ca/ycww/life-onlineStenger, T., & Coutant, A. (2011). Ces réseaux numériques dits sociaux, *Hermès*, 59, 917.

Süss, D., & Hipeli, E. (2013). *Compétences* médiatiques: Conseils pour utiliser les médias numériques en toute sécurité. Berne, Switzerland: Jeunes et médias Programme national de promotion des compétences médiatiques.

Terranova, T., & Donovan, J. (2013). Occupy social networks: The paradoxes of corporate social media for networked social movements. In G. Lovink & M. Rasch (Eds.), *Unlike us' reader: Social media monopolies and their alternatives* (pp. 296–311). Amsterdam, The Netherlands: Institute of Network Cultures.

Tisseron, S. (2011). Intimité et extimité. *Communications*, 88, 83–91.

Underwood, C., Parker, L., & Stone, L. (2013). Getting it together: Relational habitus in the emergence of digital literacies. *Learning, Media and Technology*, 38(4), 478–494.

UNESCO. (2011). *Media and Information Literacy: Curriculum for teachers*. Paris, France: UNESCO.

· 6 ·

PERCEPTIONS OF RISK AND OPPORTUNITIES

> While most children possess basic operational and safety skills, and some have also developed advanced digital competencies, they lack the maturity to reflexively engage with online risks.
>
> —CHAUDRON, 2016

Introduction

The involvement of youngsters on the Internet in general and on OSN in particular enables them to benefit from more and more opportunities but it also increases their risk exposure (Livingstone, 2014). The activities at the origin of these online opportunities (involving more or less risk) do not necessarily imply a negative effect on the users (Livingstone, Haddon, Görzig, & Ólafsson, 2011), but OSN are usually seen in two ways—one positive and optimistic and one negative and pessimistic—and there is no balanced perspective to consider both the positive and negative aspects (Fuchs, 2014).

The mistrust of OSN is high but does not prevent them from being used massively, even by adults, although they often refer negatively to OSN. They are influenced by the media and public authorities, which put the emphasis on the "dangerous and uncontrollable nature of social networks," urging parents to watch their children, trying to discourage or prohibit their use (Endrizzi, 2012, p. 16).

This attitude of parents has two consequences: (1) it leads the young people who are increasingly frequent users of OSN to dissimulate their practices and to lie, further widening the gap between generations (boyd, 2008); and (2) it prevents the parents from understanding how young people use OSN, because "the key to understanding" that reality lies in "mov[ing] away from the headlines—both good and bad—and div[ing] into [the] more nuanced realities of young people" (boyd, 2015, p. 41).

Parents are concerned about the risks to which students are exposed online, and appear to be poorly informed about the situations that they come across on the Internet: 40% of parents whose children have accessed online pornography claimed that their children did not; 56% of parents whose children were troubled online said the children were not bothered; 52% of parents whose children received sexually oriented messages said that their children did not; 61% of parents whose children have had contact online with strangers reported that their children did not (Livingstone et al., 2011).

According to the project EU Kids Online (Livingstone et al., 2011), 13% of children 9–16 years were victims of online bullying; 50% of young people 11–16 years felt more confident when they were on the Internet; and 30% of them have had online contact with strangers, although only 9% have personally met people they knew online. And if the concern for children and young people focuses on pornography in sexting (sending/receiving sexually oriented messages), on bullying, and on meeting with strangers, children are more concerned about issues such as online hate, pro-anorexia messages, self-mutilation, drug abuse, or even suicide (Livingstone et al., 2011).

On the other hand, if the concern with online risks is higher, the truth is that children and young people are exposed online and offline, as "the virtual world has not eclipsed the physical one in this respect" and the children's world, although increasingly mediated by online communication, continues to be marked by home, school, and community (Haddon & Livingstone, 2014, p. 26).

From 2011 to 2014 the percentage of young people who considered being victims of bullying increased offline (from 7% to 12%) and fell online (from 14% to 11%). Bullying offline and online varies between genders, as it declined slightly among boys (from 21% to 19%) and increased among girls (from 21% to 26%), especially in adolescents and in the age group of 13–14 years. With regard to Portugal, the cases of offline bullying grew from 2% to 5%, and the total percentage (bullying offline and online) increased from 9% to 10%. Therefore, there was a reduction in online bullying of 7% to 5%, while offline bullying increased (Mascheroni & Cuman, 2014, p. 27).

The results of the *EU Kids Online* project suggest that the "online bullying represents a new form of a long-established childhood problem rather than, simply, the consequence of a new technology" (Mascheroni & Cuman, 2014, p. 26). As an example, if the Internet is the main means of access to pornography, this access is not a new issue. The novelty is that "seeing sexual materials offline increases the odds of seeing sexual materials online by a factor of 15" (Mascheroni & Cuman, 2014, p. 29). More than half of online bullying victims, online bullies, and the ones who accessed online pornography, mentioned that they did it offline. And the youngsters who were more resilient offline also tended to be so online, though here under the influence of socio-demographic and psychological variables.

However, being bothered online (and not only being cyberbullied) is an increasing activity: 17% of Internet users said they had felt uncomfortable on the Internet within the previous 12 months. The percentages of those who felt uneasy increased with age (from 11% at 9–10 years to 23% at 15–16 years) and were higher among girls (21%) than boys (14%). Comparing the results of this project with the results of another study, *EU Kids Online*, it turns out that exposure to negative user-generated content "[has] increased substantially" (especially hate messages and those pro-anorexia or pro-bulimia), cyber bullying, and the number of offline meetings with people they met online (Mascheroni & Cuman, 2014, p. 26).

35% of Portuguese Youngster Were Harassed on OSN

In the study we developed in Portugal, we focused more on collecting troubling perceptions in OSN by young people, and less on the content to which they have access.

Answering the question, "Have you ever felt bothered by texts, photos, or videos on social networks?" one in three young people (35%, 173 people) said they had been disturbed. We did not define the period in which the nuisance might have occurred nor did we use the term cyberbullying, since the focus interviews that followed would specifically address these issues. The value is high because the studies carried out in European countries show victimization rates of around 10%, but the range is wide, from 1% to 50% (Amado & Matos, 2015). Among European and American teens, the percentage of "cyber victims" revealed by several studies varies between 9% and 34% (Lindfors, Kaltiala-Heino, & Rimpelä, 2012). In the United States, the values found in different studies with children and youngsters vary between 9% and 35% (David-Ferdon, 2009) or between 10% and 40% (Hinduja & Patchin, 2014), whereby our results are next to the highest, but within the range.

The next question was, "If you answered 'yes,' why did you feel uncomfortable?" Young people pointed to: offensive language (49%); racist messages, mockery of disabilities, and personal insults (all with 42%); violent images (30%); or the publication of unauthorized photographs in which they appeared (27%). Negative comments in relation to their country or religion were mentioned less frequently. In "Other" reasons (11%), they pointed to attacks on their favorite clubs or sports, animal abuse, sexually oriented messages, or others that affected them personally: "Pictures of my boyfriend with me," "Comments slandering my peers/classmates," "Mocking [of] things I like," "Ugly people full of 'cockiness,'" "A photo with false content."

There were more girls (42%) than boys (29%) who felt uncomfortable. In the group of girls there were significant differences between ages, because, if, for those up to 11 years, less than 20% reported having been troubled, the percentage rose to 37% at age 12, 47% at 13/14 years, and it exceeded 77% at 15 years. The data are consistent with a Canadian study, according to which the girls agreed more than boys (82% against 63% of boys) with the statement that they may be victims of strangers who they meet online (Steeves, 2014).

Also in a European study, 17% of young people reported having been bothered online over a 12-month period, 14% of boys and 21% of girls, with the percentage increasing to 23% when considering only the youth group 13–16 years. But this was the answer to a general question, without referring to specific risk situations. When the researchers asked young people if they had experienced at least one of several specific risk situations—being bullied online or offline, receiving sexually oriented messages, seeing sexualized images, meeting people offline who they met online, seeing negative contents (related to hate, pro-anorexia, self-harm, drugs, or suicide), or other privacy risks—the percentage of those who respond affirmatively grew to 46% (44% among boys and 48% among girls), reaching 63% among 15–16 years old (Mascheroni & Cuman, 2014).

A study in Finland of 5516 young people ages 12–18 showed that there is a higher percentage of female victims of cyberbullying than boys, especially in the group of 14–15 years. Girls knew more situations of peers who have been victims (16% against 10% of boys). However, considering all the individuals, the percentages are similar between genders: 11% of girls and 10% of boys said that they had been victims at least once over the previous 12 months (Lindfors et al., 2012).

In Portugal, the percentage of young people who said they had been bothered increased from 7% to 10% between 2011 and 2014, the percentages of 7% of boys and 12% of girls (Simões, Ponte, Ferreira, Doretto, & Azevedo, 2014). But when asked about having experienced a specific risk situation, 31% of young Portuguese answered affirmatively (Mascheroni & Cuman, 2014). The discomfort was always reported by young people with access via mobile phone rather than by young people who accessed the Internet in other ways, as it stands in Portugal (at a rate of 24% against 12%).

The results are similar to those we reached, with similar trends. The nuisance situations increase with age, girls, especially the older ones, said to be troubled, and young adults involved in varied platforms people refer to more online nuisance.

What Upset Youth on OSN

The focus group interviews we organized with youngsters aged 9–16 in Portugal corroborate the questionnaire data. The expression used in our questions, "feel uncomfortable in the OSN," is more general than that of cyberbullying; it is not just about an individual or a group sending repeated messages to a particular recipient in order to attack, offend, or humiliate through ICT. In addition to these "contact risks" in which young people are "participants," we also considered "content risks" where the young man was only a receiver of either illegal, harmful, offensive, sexually oriented, racist, or hate contents (Staksrud, Haddon, Görzig, & Ólafsson, 2009, p. 18) . We grouped the cases referred to by young people into eight of the nine categories of cyberbullying organized by Amado & Matos (2015, pp. 84–85), which we present and justify after the results we found. We added a category to the "content risks," with which we began:

a) **Violent images**: the proliferation on OSN of violent images that annoy young people, especially when the victims and also offenders are young people like them:

Y6—Some time ago [we were] shown a video of a student being beaten up by five peers. The video was on Facebook and...
Y3—This is awful. For us it was a bit shocking.
Y6—Usually, in social networks, we are also used to seeing [it]. There is a lot of it. We have seen a lot of violence. Basically, social networking is to show violence. So we are already used to [it]. (Excerpt from the interview with a group of 11th grade students, 15/16).

b) **Insulting** (harassment, bashing, direct attacks, verbally or with images): "I have been sent certain publications. These are people who create fake profiles and then we begin to be bothered; [they are] saying things that make us feel bad" (Rita, 15). Insults arise also in the comments to publications that lead young people to take risks, because they want that feedback. The "nastiest comments" (António, 15) that boys see/confirm happen to "girls who thus have more disgusting photos" (Luis, 13), which girls recognize, though sometimes devalue, because they are a desired nuisance: "No doubt that it exists. But the fact that girls are more troubled, I think it is because the boys are very flattering, and say 'You are very comely,' 'You are very cute.' This often happens to me" (Vanessa, 14).

c) **Heated discussion** (flaming, online discussion with aggressive and vulgar language between partners in a situation of power balance), as reported:
Y5—When [there] are games between Sporting and Benfica, for more than one day, [they] just say nonsense. It is just offensive language. Those who lose are sulking. Then blame the referee. Then they start arguing with each other...
Y4—Or comments that lead to violence.
Y6—Similar to the lists [for election to the student union] in the [school name], on Facebook were only moods and reactions [like] they post on Twitter. It was horrible, to call each other ugly names.

d) **Disclose secrets/blackmail** (outing, sharing online information or embarrassing pictures with someone, or content likely to humiliate the person, for material, sexual, or other benefits): "I have a friend [who] a boy asked to take a naked picture, and he would do I do not know what. She took and sent him a picture...with the breasts on display. He got the picture. He did nothing. But in the following year, when she had a boyfriend, he sent the picture to her boyfriend on a social network. Everyone saw and started kidding her in school" (Inês, 13).

e) **Slandering/denigrating** (denigration, putting down, misinformation, "slag[ging] off someone" on the web): Young people under 12 said they created fake profiles because "it is funny mocking people" (Madalena, 12), "[to] troll a person without her realizing who is doing it" (Inês, 11), or "[to] slag off people" (Mariana, 11). Young people 13–15 also do it to "play and mock people" (Miguel, 15), "[scold] the others" (Pedro, 16), or "[tear] the others apart" (Liliana, 16).

f) **Cyber stalking** (repeated harassment that can cause fear in victims): Boys are accused by girls of bothering them, because "they make comments, send many messages, and are always mocking us" (Margarida, 12). But the boys also complain: "A girl was always commenting on my pictures, saying I was 'beautiful and cute.' She had already sent me 80 messages or so and I had to block her" (Duarte, 12).

g) **Dissemble/usurp identity** (impersonation, posing, pretending you are someone else, usurping his/her personal data, posting messages in order to create problems for the person): "Some even create profiles with photos of other people, for example, someone creates a fake Facebook as if he was a student in our class and he is not. And you begin to invent things just to hurt this student who he says he is" (Gonçalo, 12); "There are people who like to [impersonate] a public personality and make fake profiles" (Bruna, 13).

h) **Sexually oriented messages** (sexting, sending or receiving sexually oriented photographs, as partial or full nude): "I have already denounced several Facebooks of others for sending me porn" (Helena, 14). And not only in messages, because "it is the reality I see...a lot of time, people [are] posting pornography. Nowadays there is increasingly [more] pornography on Facebook" (Marta, 13); "A classmate of ours who is no longer here asked a classmate to send nude photos of her and upset her, but she did not send them" (Luis, 14).

i) **Enticing or pretending to be a friend** (trickery; persuading someone to reveal embarrassing information): The situation will be recurrent, so that young people face it naturally:

   Y5—One was a little boy, who is from here, and I stopped talking to him, it was [name used in the profile], a fake profile. And I knew his true profile, without him knowing. He was talking to me and thought I did not know who he was, but I knew who he was.

   INT—But how did you know who he was?

> Y5—Through friends' information, contacts. That boy started [to] upset me, I do not know what, "You're very beautiful," "You have something...do not know what." But he told it to many others. And I stopped talking to him and [did] not even [answer] him.

Young people show special attention to the profiles when they receive friend requests (to use the language of Facebook), mainly due to the many existing fake profiles. But they consider that "it is…very easy to find them," because "the person is very beautiful and says [they are] very social but [they] do not have photos with friends and it is called fake" (Juliana, 13), and "as they always post photos of very beautiful people, the gang distrust immediately" (Rita, 16). In the fake profiles "the person has the same photo for a long time and publishes strange things" (Elisa, 16). Another strategy is to analyze the profile: "We see friends in common. If we do not have any friends in common, how can [they] be asking me to be friends?" (Francisco, 16). The identification of who is beyond the fake profiles is a result of different information…the interpretation of the very fake profile, publications associated with it, and the relations established on the OSN, as the profile always reveals something about its author:

> We can damage the platform by publishing a photograph of another person as our profile photo, [beside] a false name; we may decide to represent us [by] other types of images, of our domestic animals, a landscape, of our family. But we cannot get away from some self-representation. (Enli & Thumin, 2012, p. 101)

These profiles can be considered to be the base of a third category of risk, the "behavior," in which the young man is an actor, and this is also an example of the provision of personal information about third parties, downloading or illegal gambling or illegal access to computers or accounts (Staksrud et al., 2009, p. 18). But not all the fake profiles are intended to annoy someone. Some are created with other purposes, which may include overcoming shyness or prohibition of use of OSN to obtain benefits in online games or in terms of popularity.

Fake Profiles: Innocence vs. Purposeful Harassment

On OSN, users are identified by their profiles as they only come in with login and password, which links the user profile and all interaction developed through it. "[To] try to escape this identification, many users choose to create fake profiles and use them for interactions in which [they] do not wish to be recognized by the others" (Recuero, 2009, p. 28). These profiles are often referred to by young people, some of whom admit to having them, and they exist in all OSN, being created for specific purposes, such as:

a) **Interacting anonymously**: "In ASKfm, as you cannot ask anonymous questions, I had an Ask where I was Mary, to ask those anonymous questions" (Juliana, 13); "[it] may be for not feeling well about themselves. There is a series about it. Often they talk to people who [they] already know but as [they] do not like their looking, [they] use fake profiles" (Marcia, 15).
b) **Talking to boys/girls** (referred by young people 13–15 years): "To see the publications of other girls and to comment without being identified" (Luis, 14); "The fake profiles are all, or almost all, created by girls. [They] take advantage. They say, 'Look, this girl [is] so cute. I am going to create a profile with her photo.' I know one who did this and spoke with boys and boys every day" (Mariana, 15).
c) **Participating in OSN without parents' knowledge** (among young people under 12 years): "I know peers who are already in other schools and [who] used Facebook without their parents' knowledge. They used fake names" (Duarte, 11).
d) **Progressing in OSN games**: "Sometimes I create fake profiles which are to connect [in] the games. You get more things if you have more people playing" (Pedro, 12); "I created five fake accounts because of the games, to win, to send game stuff to myself, and so I got stronger" (António, 15).
e) **"Liking" and posting fake comments**: The "false likes" (Rui, 15) can be made in [our] own publications or [in] others'.

Despite these reasons, the most common situation is the association of fake profiles with the annoyance of people on OSN. The terms bullying and cyberbullying were even introduced in interviews at students' initiative, for example, when they said that "there are cases of bullying because of fake profiles" (Paulo, 15). Using this terminology or not, young people, teachers, and parents react to nuisances on OSN.

Youth and Adults React to Online Harassment

We asked the young people to refer to their coping strategies, that is, how did they react when they felt uncomfortable on OSN ("What did you do when this happened?") The most common reactions were skipping [the messages] (61%) or blocking (50%) the author of the content who generated the nuisance (Figure 1).

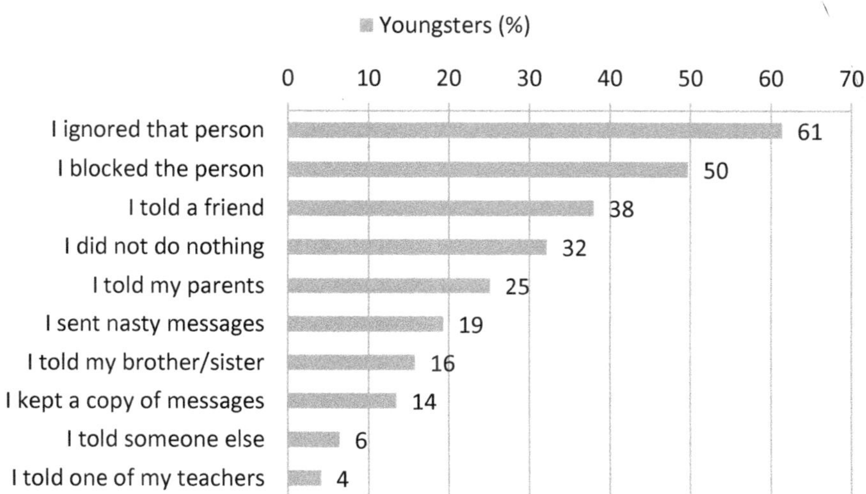

Figure 1. Reactions of young people in situations of nuisance in the OSN (%) N = 171.

Survey responses to nuisance situations fall within the five groups of standard replies organized by Mascheroni and Cuman (2014). These responses are not mutually exclusive since young people sometimes use more than one of them to cope with the same problem.

a) **Follow prevention strategies** such as self-monitoring to avoid risks, also adapting the privacy settings: They avoided, for example, more intimate contact with strangers in order to protect themselves:
For example, I [found] a boy on Facebook and I [talked] to him. And then we started talking, we talked every day sending photos to each other and [I] do not know what. Who gives a lot of confidence to these people; it then comes to a point that [it] does not work. They can enter into a relationship that does not work, because they do not know [each other] well. Then it happens that [we] sent photos to each other...and then when they get bored, the pictures appear on the Internet. (Mariana, 15)
They also adjusted the privacy settings:

They bother us. It is not only contents, but also sometimes with comments. But it also has to do with the privacy of our account. If we do not organize the privacy and let the profile [be] universal, it is clear that [it] will bring problems for several people. (Daniela, 14)

b) **Technical measures** imply an interaction with the equipment used to access or with the service and thus knowledge to do it: Half of the respondents agree to having blocked the person or entity that caused the nuisance and 14% saved a copy of the message that annoyed them. The action of blocking (sometimes reversed when students have tiffs and then solve the issue) is the most reported, mainly as a result of the excess of incoming messages by the other person, or unwanted comments or shares in your profile, regarding Facebook: "There are people who make a lot of comments, send many messages, and are always mocking us" (Margarida, 12). The excess of notifications received can also lead to blocking people: "Sometimes on Facebook we have some 20 notifications asking that we share this or that, for example, how to change the color of the background and such things"..."or shares that have to do with games" (Francisco, 15).
c) **Confrontation** with who bothers him/her sometimes, to clarify and prevent the issue from getting worse, although the conflict is common: A fifth (33 individuals) responded with nasty messages, looking to bother who bothered them, something mentioned in the focus interviews: "When a person does something to me, I comment saying bad things about it" (Marco, 11). When asked whether they were messages or publications authors aimed at disturbing others, the situation changes. To the question, "Have you ever sent text, photos, and/or videos with the intent of bothering people?" 492 of the 497 young people answered, but only 18 (4%), 15 of them boys, admitted it. In terms of confrontation there was a combination of answers to solve the same problem: "I complained to her because she was always sharing to my profile and I did not like [it]. As she did not do anything, I deleted her [blocked her]" (Gonçalo, 12).
d) **Doing nothing** is the reaction when they consider that the technical assistance does not solve the problem, nor the help of others because they believe that adults will not be interested in the question, they do not take them seriously, and can get angry, scolding them or punishing them: A third of young people (32%, 55 individuals) said that they had done nothing, which is consistent with a European study's

results in which one in three young people said they did not talk to anyone about what bothered them on the Internet (Mascheroni & Cuman, 2014). However, in our study and because it was possible for respondents to choose more than one option, we found that only 24 of the 55 individuals chose the option "I have done nothing." So 31 of those who said they had done nothing, did something more, as they also pointed out the following: "I ignored that person" (24 cases), "I told a friend" (16 cases), "I told my parents" (12 cases), and even "I blocked the person who sent me messages" (7 cases). The fact that they did not confront the offender may have been understood by young people as having done nothing.

e) **Collective approaches** are common, as young people realise that the support of people they know is important to deal with unpleasant situations experienced online, either on the emotional or technical level. They are mostly mentioned in cases of bullying: Young people said they had told what happened to others, friends being their first choice (38%), followed by parents (25%), siblings (16%), or others (6% reported family, best friends, or lovers), and then their teachers (4%). The percentages we have reached are lower than those of *Net Children Go Mobile*, where 67% of youngsters (61% of boys and 73% girls) reported talking to someone about situations in which they felt uncomfortable on the Internet, in particular with mothers (71%), friends (57%), or parents (54%) (Mascheroni & Cuman, 2014). They are not close either to the data of the integrated Portuguese investigation on that project, in which 80% of young people admitted to speaking to someone in cases of online nuisance, parents being the first choice (68% the mother and 53% the father), followed by siblings (36%), friends (32%), and teachers (20%). "When compared to the rest of the countries, the Portuguese children [mentioned more that they] seek support from someone (80%), for an average of 67% and a minimum of 56% in Denmark" (Simões et al., 2014, p. 36). But there is a difference between the studies because while we asked what they did when it happened, in the referred studies the people were asked what they would do if the nuisance situation happened. And research in this area has shown that when children are asked to whom they should turn if they are victims, they first say their parents, then teachers and police, and then peers/classmates. But "when you ask them what they did the results are very different: 35.9% of the victims asked for help [from] friends, 27% [from] parents, and 9.3% [from] teachers" (Amado & Matos, 2015, p. 95).

Only 25% of young people who felt uncomfortable while on OSN talked about it with parents and only 4% told teachers. But young people, teachers, and parents reported cases where the action or the will to act was clearly on the part of teachers and parents. One of those cases concerned a girl who met someone she had met online in person:

> INT—Teachers care about you in this issue of [being] online and [the] Internet. It seems they know more about you than your parents...
> Y4—Teachers go through many students and learn from these people. Our parents only live with us. And when there is news about it [they] soon become very concerned.
> Y7—Yeah, there was a time in our class of the 9th grade, the [girl's name] spoke to our class director, remember? She said she spoke with someone on Skype, who [she] has never seen. And then she went to Lisbon to meet her. I remember our teacher scolding her very, very, very much... scolded a lot, a lot. The [girl's name] said she was with her uncle and they used the webcam, but the teacher said that she should not have gone, because something could have happened. The truth is that her grandmother, with whom she lived, and his uncle, they also allowed her to go.

Teachers also reported cases such as one between a student's parent and the school. The teacher interviewed believed that parents should be more attentive to what students do on OSN:

> T—They comment much about the photos they place there [in the OSN]. And contact a lot through Facebook. Once I even told a girl's father of seventh grade: "If I had a daughter your age, I would not allow [it]." It is complicated afterwards. Because there are less proper comments [from] students/peers about the photos the girl publishes.
> INT—But do they choose more provocative photos?
> T—They are already grown up girls. They put on airs. And they comment. And there are bad and less bad comments. Some parents had already come here complaining. But then, where does the power of [the] school end? There are parents who complain about the girl's classmates. It already happened.
> INT—Therefore, parents assume that the problem is the effect and not the cause...
> T—Well. The act of posting photos should be controlled. But no. Parents do not control things.

The parents requested that the school take action. This also occurred in other cases, such as when students were photographed in awkward situations and those pictures were published on OSN:

T—This year, in the 10th grade there was a problem. A father came here to complain that someone published his child's photo during a lesson out there [in the courtyard outside the school]. And then he came here saying he did not want the photograph there because his peers have commented. The father came to talk to the class director. He wanted to know why they did it and how was it possible.
INT—But were the comments aggressive?
T—They were a bit nasty. And it was a boy. Not a girl.
INT—But was it because of the way he was dressed?
T—No. It is a professional course. Sometimes they do field classes. And at the back there are a few cacti which give some figs. The boy began eating a fig and they took a photo of him eating figs. And the father did really want to find out who took the photo. The issue here, again, is: how far is it going [to become the] when does it become the school's responsibility? I do not know how it ended, but I know the photo was taken out of the classroom. But it was a delicate matter because they were in class. The teacher did not notice, and suddenly, it was already on Facebook. Then they [asked the] school [to take responsibility for it]. I do not know if this is legislated. I know there is invasion of privacy.

Parents' and teachers' concern regarding the use of OSN by young people is high, as is made clear when they are asked to assess the risk of a number of actions that could be developed by young people on OSN. This comes from the common view that children and young people are "vulnerable" on OSN. But going beyond this simple view, it is certainly important to empower them in the digital world, according to Livingstone (2014), following an exploratory study in which young people (9–16 years) of four European countries (Spain, the UK, the Czech Republic, and Romania) were interviewed. This study aimed to understand the relationship that young people establish between opportunities and risks online, how they learn to interpret what are the risks and opportunities, and opportunities that involve risk ("risky opportunities"), why do they engage in risky situations and with what implications? In short, how do they develop social media literacy?

In this exploratory study the researcher stated that young people's skills, opportunities, and risks are correlated, so those who are the most competent in the technical and social use of OSN most engage in risky situations. So aiming to develop young people's skills in relation to social media to benefit from the opportunities and avoid the risks is somewhat paradoxical. Perhaps seeking to understand what children and young people already know, what social media literacy levels they have reached, and how, will help to realize what should be done in terms of media education and educational intervention, in order to avoid risks.

The study results show that young people, regardless of age group (they were organized in levels 10–11, 11–13, and 14–16 years), are receptive to talking about OSN and the associated risks, and many experiences are reported by the classmates/peers of those who lived them or they refer to cases experienced through the media. But when it comes to bullying, pornography, hacking, or violence cases they report in the first person, which matches our practice on the topic. They pointed to the differences in the age of young people and the social context into which young people are inserted, which is also found in our study.

To Summarize, a third of the youngsters stated that they have felt uncomfortable on OSN, especially girls, whose reported cases increased with age. Complaints related to pornography, sexting, insults, and violence, many of which originated from fake profiles that young people already face naturally, to the point of identifying some authors of those profiles. Young people react with strategies "considered preventive and others reactive in nature" (Amado & Matos, 2015, p. 94), especially blocking the authors of the discomfort confronting them or complaining about it to close adults and peers. In effect, they demonstrate certain levels of social media literacy. This area needs further exploration in order to understand the perception that youngsters have of the opportunities that involve risks, how they learn to deal with them, and the consequences of these situations. Studying youth social media literacy will be important in terms of defining activities for media education strategies to develop with youth.

OSN: Young Users/Fearing Adults

In the quantitative phase of the study we asked teachers and parents to assess 10 common actions of young people on OSN, offering four choices: "low risk," "medium risk," "high risk," or "do not know." Figure 2 includes only the response rates for the option "high risk," plus the percentage of non-response, and enables us to realize that if the majority of parents assign high risk to the actions, the percentage of teachers who do so is even more.

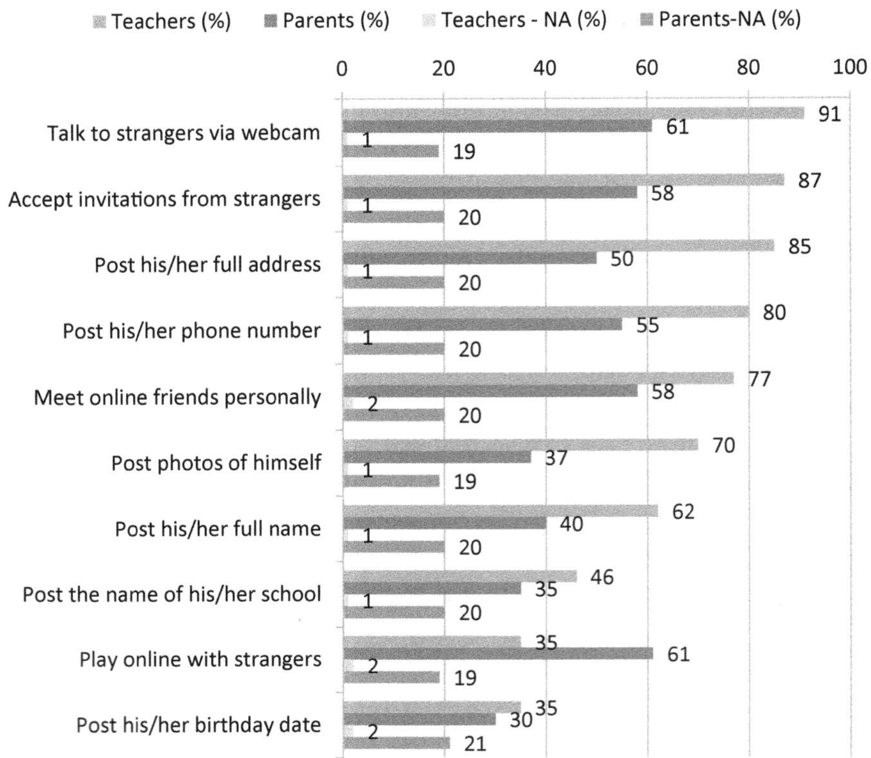

Figure 2. Risk assessment of young people's actions on the network by teachers and parents (%)N = 150 (teachers); 267 (parents).

Using a webcam to talk to strangers was the action that the largest percentage of teachers (91%) and parents (61%) considered high risk. But also the relationship of young people with unknown people worried teachers and parents, whether it be receiving invitations, meeting online friends in person, or even playing online with virtual friends.

The publication of personal information generated varied levels of concern: revealing one's address or phone number was a higher risk activity for more teachers and parents than publishing photos of oneself, the name of the school, or even one's full name.

The percentage of teachers who considered those to be high risk was always higher than that of parents by about 30 points, which was attributed, in interviews, to a greater perception of danger by teachers and/or a lack of knowledge of danger by parents. The answer to the item "play online with strangers" demonstrates this, as it is the second that most parents (61%) considered high risk, but one that fewer teachers (35%) ranked this way. Parents interviewed considered that "parents have not understood [it] well because they do not play online. Parents do not realize they are playing with a friend from elsewhere, they do not know" (P7). Teachers know that online games are often played with strangers, but parents do not know this:

> I play quite often and there is no problem. The moment you hear the word "unknown," you connect with "dangerous" straight away. In a game I do not even need to identify or know where I am. It is the "unknown" word that scares [me]. (P1)

From the risk assessment made by teachers and parents, we organized individuals into five sections: "Very Concerned," "Concerned," "Balanced," "Permissive," and "Do not know to assess/no answer." The profiles have been defined from the sums of each individual response. The answer "Do not know" was identified with 0, while "Low risk" was 1, "Medium risk" was 2, and "High risk" was 3. Therefore, those who classified all the activities as high risk activities would have a sum of 30. We considered individuals with the sum equal to, or less than, 15, "Permissive"; those between 16 and 20, "Balanced"; those with responses between 21 and 25, "Concerned," and those with a sum not exceeding 26, "Very concerned" (Figure 3).

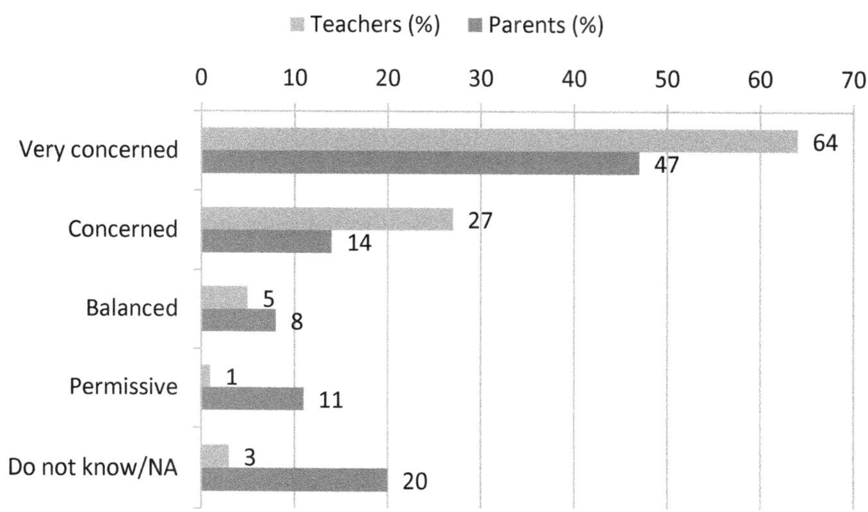

Figure 3. Profiles of teachers and parents regarding risk assessment (%) N = 150 (teachers); 267 (parents).

Two thirds of teachers and half of parents were "Very Concerned" about the practice of these actions by their students/children on OSN. One in four teachers said they were "Concerned," while 15% of parents felt differently. If we interpret the "Concerned" and "Very Concerned," we found that 91% of teachers and 61% of parents considered the actions evaluated mostly medium or high risk.

In all profiles, "Balanced" and "Permissive," the percentage of teachers is only 6% and that of the parents is almost 20%, which means that there is a higher percentage of parents than teachers that considered the actions medium or low risk.

Teachers agreed with parents and went even further, stating that all the parents' answers would have a considerable margin of error: "I believe that many of these parents do not know what they are talking about" (P1); "Perhaps the parents do not realize, they are not informed...they spend little time with their children" (P2); "[They] do not know what their children do on the network" (P3), so "they might not have answered very consciously" (P24), and there are some cases that teachers consider extreme:

> There will probably be parents who do not know, for example, that you can chat with strangers in real time, through video. This has happened to me. Once we were talking about video surveillance cameras, at school, and a father did not believe [it]. He told me it was only in the movies. And I took him to the office where the monitors are. He saw and then told me he thought it was only in the movies. (P7)

All respondent teachers recognized that they must be better prepared and more alert than parents to the dangers that young people are subject to on OSN. They confirmed parents' perceptions at two levels:

a) Teachers are doubly concerned, by also being parents: "I am a teacher and have children. As a mother, I would answer as teachers did... .I do not understand how parents answered that way" (T13).
b) More protective than parents: "Teachers are more fearful because they know better what they are talking about" (T14).

Teachers thought that they had more information than parents, but it is "common sense" information (T10), though more consistent and frequent "because teachers know of more cases where there were problems than parents...we have access to a range of reports... [because] we get in touch with many young people, while [for] parents [it] is more with their child or children"

(T25). Teachers "hear the community" (P31), "hear talks among students" (T2), are "more alert" (T32), and more "easily aware of the risks associated with these behaviors than parents" (P7), since "the students, even the shyest, keep talking with us, but will not do so with their parents, alone, without peers/classmates around" (T8).

Overall, teachers thought that they were better informed than parents about the risks that young people are exposed to on the Internet in general and on OSN in particular, resulting from their contact with students and awareness of cases. They stated that they were more concerned than parents, as they do not know what the students do, in fact, on OSN. Parents agreed that they pay for the students' access to the Internet, but they allow the students to use the Internet and OSN without monitoring (De Abreu, 2011), although there are those who claimed not to understand the lack of knowledge on the part of parents to assess the students' risk of actions on OSN, because "today there is a great concern to inform parents about the dangers on the Internet. You only do not know the dangers if you do not want to" (P18). That was one of the questions posed to parents and teachers in the questionnaire, the analyses of which are as follows.

Concern Increases With Information on Risks or Not?

We asked teachers and parents to what extent were they aware of the risks that young people are exposed to when using OSN. Most parents (67%) thought that they have much or all the information, and that percentage was strengthened among teachers (85%). Only 13% of parents and teachers admitted to having only some or no information, and 20% of parents did not respond.

But when asked if they resorted to looking up Internet addresses or other services about the risks to which young people are exposed on the Internet in general and on OSN in particular, 67% of teachers and 65% of parents said no. Only 31% of teachers and 16% of parents declared that they resorted to one or some services, which leaves doubts about the fact that the vast majority of teachers and parents thought that they have much or all information, since they did not say where they got their information from.

Teachers were more cautious in interviews, saying that "today we cannot say we have all or much information, because things evolve very quickly" (T13). Then, "just answering that we know everything…is a problem" (P40) as "teachers perhaps think they know, but do not know" (T14), which is materialized with examples: "If I look at the entire population of my school, most people have never heard of SeguraNet [Safety Net], [they do] not even know what it is. A lot of people know, but most do not have the slightest idea" (T19).

Our results showed that teachers who felt that they have much or all of the information did not resort to looking into Internet sites or support services, so they might have found it other ways. Teachers pointed to sources they use but did not report on "common sense" (T1) on "magazines, newspapers, and television" (T25), on "Internet and others who are closer or immediate" (T13), and on "news about scientific studies in this area" (T2). They highlighted "training in schools done by several entities, such as SeguraNet" (T25).

Parents considered that "no one has all the information" (P17), although there is a larger knowledge of parents resulting from "what they learn on television and hear others talk [about]" (P12) and what they "learn from each other" (P3). They valued life experience because "many of the Internet dangers are everywhere" (P18), so "it is a matter of common sense" (P19).

The interviews do not confirm the questionnaire answers, although most teachers and parents claimed to have lots of information and all the necessary information, the reports of respondents tended to discredit that point. As a teacher said, "if a third of the teachers know where to seek information on the Internet, it is very good" (T7).

Information to Help Keep Students Safe

We asked teachers and parents if they would like to have access to more information about Internet safety, and if so, what support, and how would they prefer to access it? Of those that responded, all claimed to be interested in receiving more information, especially in traditional media formats such as e-mail and printed documents.

Receiving information online through a website or newsletter was an option for one fourth of the teachers, but less desired by parents, and about 10% of individuals would like to receive information in a downloadable format. The teachers who resorted to Internet sites to obtain information on security stated that, further, they wished to receive information that was downloadable from the Internet. This option was referred to by 44% of teachers who had resorted to websites and only 19% of those who had not.

In the interviews, we addressed the reasons why they wanted to receive further information and their preferences for receiving it in traditional media. Teachers justified this with: "If they receive [the information] they do not have to look for it" (T2) and said that many of them "feel they should have further information because they feel that students are ahead of them" (P7).

Parents mentioned issues of access, because "in the computers they take too long; if they get the information near at hand [it] is easier" (P5), and often mentioned the need to be aware of developments in technology "to keep pace with children" (P17).

Teachers who preferred online information considered that "information is the same, whether digital or printed" (T14) and highlighted the advantages in terms of access, portability, and the fact that it does not take up space at home. But they accepted that "most even prefer the traditional" (T14) because they "are still too conservative" (T25), "are not used to [going] online" (P4), "do not know quite well where to look on the Internet" (T5), and "suffer from paper addiction" (T8), as "if you send an e-mail to teachers, they will print [it] because they like to underline and take notes" (T8).

In a school where all notices and even mail, which is previously scanned, are sent by e-mail, the option did not reduce the printing on paper, as realized by the dean: "When I come to the staff room, I realize they are printing the [e-]mails, because they need to read on paper" (T7). The tendency to print is related to habits, but they are also afraid of losing information:

> Teachers got used to fil[ing] so, when they have important information in the [e-]mail, the first thing they do is print and save. Especially because many of them have had virus or disk problems, or they already lost a pen. They fear losing information online. So I tell them to use Dropbox. But they still fear it. (T24)

The association between printed text and credibility also occurs with parents because the printed information "is more secure, more reliable" and has "credible sources" since, "for example, PSP itself [Public Security Police] often says that it does not send information or warnings via e-mail or the Internet" (P3).

In conclusion, teachers and parents have less information than they claim to have about security information using the Internet. The vast majority cannot say how they acquired the information they claim to have; the vast majority want to get more information, to keep up-to-date, and, preferably, by mail and in print, as they tend to rely more on print and have the habit of reading on paper.

Conclusions

Adults were more focused on the risks than on the opportunities of OSN, as 91% of teachers and 61% of parents were "Concerned" or "Very Concerned" about the risks to which young people are exposed, focusing their action to avoid the risks and not to take advantage of the opportunities they offer (Gee, 2010).

Most parents and teachers thought that they were informed about those risks and their consequences, but had some difficulty explaining where the information came from, showing in the interviews that their main sources of information were common sense and the media. Therefore, teachers and parents would be subject to "myths" associated with "real incidents" or "data that have been magnified disproportionately or are exaggerated deliberately to trigger fear," since "media culture exaggerates this dynamic, expanding anguish and strengthening fears" (boyd, 2015, p. 34).

Most teachers and parents considered that teachers are better informed about the risks to which young people are exposed on the Internet because they deal with them and know more cases. But the results do not prove that increased information because the sources of such information, in addition to the contact with young people, are similar.

Teachers and parents said that they were receptive to more information about the risks associated with participation in the Internet and on OSN, though they would prefer to receive it in traditional formats such as e-mail or printed. This availability of more information must be supported from schools and other entities, as 65% of teachers and 76% of parents admitted talking about OSN with their students, therefore they must be prepared to "strengthen the capacity and emerging competencies" (Jenkins, 2009, p. 114) of children and young people early on, and in a way that is not to prohibit the use, but prepare them for the active and practical use of OSN (De Abreu, 2011).

As to whether youngsters or adults develop strategies to avoid the risks they face when using OSN, they monitor their profiles and the comments made by third parties on their publications, they are particularly careful with passwords, and they conceal personal information in their profiles. Such care is higher among females, whether they are youth, teachers, or parents, showing in the results that there is a difference in terms of gender. But monitoring has two sides: some monitor because they have reached digital literacy levels that enable them to handle profile management; others protect themselves from failure, not actively participating in the OSN and not publishing personal information. That has produced a deficit in terms of self-representation, an essential condition to being visible on OSN and seizing the opportunities they offer (Enli & Thumin, 2012). This attitude is less common among young people, who show more of themselves to create new relationships and expand the circle of friends (Livingstone, 2014) on OSN.

Young people also have strategies of self-protection in the OSN, in particular, risk prevention (e.g., avoiding contact with strangers), the implementation of technical measures (e.g., blocking those who bother them), confrontation (e.g., harassing those who bother them), or collectively approaching problems (e.g., reporting troubling situations to parents, peers, or teachers). These techniques, which are not mutually exclusive, indicate that young people have reached certain levels of social media literacy, whose acquisition process should be better studied and understood (Livingstone, 2014).

But that does not mean they are ready for risk situations, as one in three young people claimed to have been troubled on OSN, ie., the girls (e.g., violent images, insults, discussions, blackmail, slander, cyber stalking, identity theft, enticement, or sexually explicit messages). Some cases were known to parents and teachers who feared for what might happen to their students, warning them of the dangers and/or imposing restrictions on the use of OSN, actions that are likely to be revisited and reviewed as young people progress in terms of social media literacy. As young people report more to peers than to parents or teachers, whether or not they are victims online, "a major investment must be made in children and youth education in order to develop skills to help peers/classmates/friends involved in cyberbullying" (Amado & Matos, 2015, p. 96).

References

Amado, J., & Matos, A. (2015). O *cyberbullying* entre os comportamentos de risco *online*. In Guilhemirna Miranda (Ed.), *Psicologia dos comportamentos online*, Lisboa, Portugal: Relógio D'Água Editores.

boyd, d. (2008). *Taken out of context: American teen sociality in networked publics* (Doctoral thesis). Berkeley, University of California.

boyd, d. (2015). *É complicado—as vidas sociais dos adolescentes em rede*. Lisboa, Portugal: Relógio d'Água.

Chaudron, S. (2015). Young Children (0–8) and digital technology. European Commission.

David-Ferdon, C. (2009). *Hertz MF electronic media and youth violence: A CDC issue brief for researchers*. Atlanta, GA: Centers for Disease Control.

De Abreu, B. (2011). *Media literacy, social networking and the Web 2.0 environment for the K–12 educator*. New York, NY: Peter Lang.

Endrizzi, L. (2012). Jeunesse 2.0: Les pratiques rélationelles au Coeur des médias sociaux. *Dossier d'actualité Veille et Analyse*, *71*,1–24.

Enli, G., & Thumin, N. (2012). Socializing and self-representation online: Exploring Facebook. *Observatório*, 6(1), 87–105.

Fuchs, C. (2014). *Social media: A critical introduction*. London, England: Sage.

Gee, J. (2010). *New digital media and learning as an emerging area and "worked examples" as one way forward*. New York, NY: MIT Press.

Haddon, L., & Livingstone, S. (2014). The relationship between offline and online risks. In C. von Felitzen & J. Stenersen (Eds.), *Young people, media and health: Risks and rights*. Gothenburg, Sweden: Nordicom.

Hinduja, S., & Patchin, J. (2014). *Cyberbullying: Identification, prevention, & response*. Retrieved from Cyberbullying Research Center website: http://cyberbullying.org/Cyberbullying-Identification-Prevention-Response.pdf

Jenkins, H. (2009). *Confronting the challenges of participatory culture: Media education for the 21st century*. New York, NY: MIT Press.

Lindfors, P., Kaltiala-Heino, R., & Rimpelä, A. (2012). Cyberbullying among Finnish adolescents—a population-based study. *BMC Public Health*, *12*, 1027.
Retrieved from http://www.biomedcentral.com/1471-2458/12/1027

Livingstone, S. (2014). Developing social media literacy: How children learn to interpret risky opportunities on social network sites. *Communications*, 39(3), 283–303.

Livingstone, S., Haddon, L., Görzig, A., & Ólafsson, K. (2011). *Risks and safety on the Internet: The perspective of European children: Summary*. London, England: EU Kids Online.
Retrieved from http://eprints.lse.ac.uk/33731/

Mascheroni, G., & Cuman, A. (2014). *Net children go mobile: Final report*. Retrieved from http://netchildrengomobile.eu/ncgm/wp-content/uploads/2013/07/NCGM_FinalReport_Country_DEF.pdf

Recuero, R. (2009). *Redes sociais na Internet*. Porto Alegre, Portugal: Sulina.

Simões, J., Ponte, C., Ferreira, E., Doretto, J., & Azevedo. C. (2014). *Crianças e meios digitais móveis* em Portugal: Resultados *nacionais do Projeto Net Children Go Mobile*. Lisboa, Portugal: CESNOVA.

Staksrud, E., Livingstone, S., Haddon, L., & Ólafsson, K. (2009). *What do we know about children's use of online technologies?: A report on data availability and research gaps in Europe*. London, England: EU Kids Online, Deliverable D1.1.

Retrieved from http://eprints.lse.ac.uk/24367/1/What%20do%20we%20know%20about%20children%E2%80%99s%20use%20of%20online%20technologies(lsero).pdf

Steeves, V. (2014). *Young Canadians in a wired world, phase III: Life online*. Ottawa, Canada: Media Smarts. Retrieved from http://mediasmarts.ca/ycww/life-online

· 7 ·

LEARNING PERSPECTIVES FROM STUDENTS, TEACHERS, AND PARENTS

> With social media use rapidly growing across Europe, there is a need for schools to develop policies and guidelines so that students can better and more safely leverage these platforms and tools for learning.
>
> —JOHNSON ET AL, 2014

Introduction

Citizens, younger ones in particular, "learn through their practices in social networks," but the educational and training institutions take advantage only marginally of the opportunities offered by OSN (Endrizzi, 2012, p. 12). Some are afraid that something bad might happen, because they are frightened by the excessive attention that the media confer on the most negative cases. Others have not yet realized its full potential in terms of education (Gallagher, 2011). Meanwhile, young people gradually improve their ability to use social media and show a clear "preference for interactive learning outside school" (Underwood, Parker, & Stone, 2013, pp. 479–480), by questioning the traditional monopoly of knowledge assigned to it.

It is urgent to change this attitude because social media, including OSN, and its massive use, continues to grow "new opportunities for innovating and modernizing Education and Training institutions and for preparing learners for the 21st century," as stated Redecker, Ala-Mukta, & Punie (2010, p. 4), sustained in two studies, one that analyzed the potential of social media in formal learning and another that examined the learning communities and networks of informal learning. The studies found that social media open innovation possibilities in both contexts because they:

a) Provide access to a wide variety of content (often free) that supports learning and professional development throughout life, contributes to equity and inclusion, and puts pressure on the school and other institutions to improve the quality and availability of its learning resources;
b) allow users to create their own digital content and publish online ("user-generated content"), creating conditions so that everyone can benefit from each other with such content, encouraging a proactive approach to learning;
c) connect learners with each other, but also with experts and teachers, enabling the validation of the content they produce, but also access to specific knowledge in a given area;
d) enable learners and teachers to collaborate on projects or discussions of mutual interest issues, joining expertise and potential in a group with a common goal.

These four dimensions (content, creation, connection, and collaboration) were labeled as the four Cs learning 2.0 in research developed under the supervision of the Institute for Prospective Technological Studies (IPTS) with the year 2020 as the horizon (Redecker, Ala-Mukta, & Punie, 2010, p. 4). On the whole, they boost pedagogical innovation, changing the relationship between teachers and students by making it more personalized and creative. They boost organizational innovation, calling for a more dynamic school, flexible and open, which involves creating accessible technological infrastructures for teachers and students, an atmosphere of learning support 2.0, and new models of teaching, learning, and assessment. They still boost the integration of formal learning environments, non-formal and informal, which is understandable, because "the mediatic-educational question is not only a school issue but also a family issue and an issue for all educators who work in the territory" (Rivoltella, 2012, p. 25).

Formal learning takes place in education and training institutions (Perulli, 2009), is organized according to a program, structured, purposeful, and aims to obtain certification (Smith & Clayton, 2009). It is a closed system that operates according to a prescribed logic, based on a curriculum and pre-established roles of its agents because teachers teach and students learn by reproducing the learning in tests to obtain degrees (Underwood et al., 2013).

Informal learning occurs in everyday life and may not be recognized by the learners for being unintentional, and having a lack of structure and association with an educational institution (Underwood et al., 2013). It is an "open-negotiated system, amenable to new information, tools, tasks, and participants, as worked out in the course of ongoing activity" (Underwood et al., 2013, p. 482).

a) Formal learning is distinguished from informal learning to six levels (Jenkins, 2009);
b) the formal is more static, lasting in time, while the informal is more innovative and has no predefined temporal dimension;
c) the formal has an institutional structure, while the informal is more or less organized, according to the agents involved;
d) in the formal, community changes little over time, while in the informal, that community is dynamic and can exist only in the short term;
e) the formal is bureaucratic and usually nationwide, while in the informal, community is formed ad-hoc and is of volatile location;
f) in formal, mobility (of: school leavers) is limited, while in the informal, participants can leave the learning community at any time if, for instance, it does not correspond to their expectations.

The integration of formal and informal learning contexts aims at the integral formation of citizens, which involves questioning and reflecting on the nature of education in order to explore new types of learning experiences for students and teachers (Underwood et al., 2013). The debate has been active in Europe since the 1990s, but "we are still a long way from either a standard theory or a practical policy on linking these different types of learning" (Perulli, 2009, p. 98) and from the implications linked to it. One is the extension of the concept of literacy and its impact in the classroom.

The Enlargement of Literacy

The articulation of formal and informal learning requires the enlargement of the concept of literacy, which keeps the traditional logic, based on reading, writing, listening, and speaking, but includes "digital literacy," which is "create, work, share, socialize, research, play, collaborate, communicate and learn" (Meyers, Erickson, & Small, 2013, p. 356).

Literacy is not only a mental phenomenon but also socio cultural (Gee, 2010; Jenkins, 2009), so the learning processes "are not just individually internalized but socially distributed, with implications for pedagogy and instruction" (Underwood et al., 2013, p. 479).

Citizens participate in social and cultural groups, resulting in experiences mediated by technology, whether printed, digital, or other (Gee, 2010). Social media in general and the OSN in particular, are the basis of an "online culture" (De Abreu, 2011, p. 51) a "participatory culture" (Jenkins, 2009, p. 8) in network but also a "culture of connectivity" (Van Dijck, 2013, p. 5).

The social production of meaning is more than the sum of productions of individuals and OSN are spaces that enable this production, so the school should integrate them and develop knowledge within the collective intelligence, act in communities with different cultures, and reconcile data conflict. We cannot continue preparing individuals to solve problems individually, because it is no longer what society asks citizens (Jenkins, 2009).

Learning involves the connection of students' interests to curriculum issues, takes into account the specificities of each one, and is carried out in spaces and times out of school. "Social learning" leads citizens to develop critical thinking and multiple capabilities, including the preparation for lifelong learning. To implement it, teachers and trainers must enhance the creativity of the students and not the reproduction of contents, organize learning through project work, collaborate with students *offline* and *online* (through common multimedia platforms) and share those experiences with other teachers (De Abreu, 2011).

Twitter is a good example of an OSN that can be used by the school as a social learning platform. In a study (Delesalle and Marquié, 2014) that involved teachers using Twitter with their students but also school principals and other trainers who follow these practices, respondents indicated that the use of Twitter reinforces: (1) the methodologies of project work; (2) interactivity of the sessions; (3) peer tutoring; (4) production and creation by young people. The study also sought to understand the relationships between adults and young people and between school and family sphere through Twitter and identified four trends:

a) Teachers most appreciated students activities, including writing, that enabled them to go beyond the school intramural contacts (setting up or strengthening social bonds), which motivated young people, even the youngest, whose extra-school contacts were limited to parents, thus strengthening the ties between the school and the family.
b) The use of OSN in the classroom involves discussion of this same use and *online* presence (a child reported that if he made a mistake, he wouldn't like to have it online), which raised issues in terms of the age of access, online identity, privacy, and protection of personal data.
c) connect learners with each other, but also with experts and teachers, enabling the validation of the content they produce, but also access to specific knowledge in a given area;
d) enable learners and teachers to collaborate on projects or discussions of mutual interest issues, joining expertise and potential in a group with a common goal.

Twitter was considered a tool among many that is not enough by itself and that even points to others (blogs, interactive whiteboard, other OSN, websites, school platforms...) that teachers may experience, also by influence of other colleagues.

Teachers no longer work in isolation, because on Twitter, or other platforms (e.g., blogs, Facebook), they manage to get training, be aware of what is going on, contact parents, share, and benefit from sharing educational resources, outside school and on online school platforms.

Among the teachers who use OSN we are witnessing a double movement: there are more and more ways to use each OSN, while there is a multiplication of the OSN used. Taking into account their potential use for teaching purposes and following the organization proposed by LeDeuff (2011), OSN may be of four types:

a) Socio-digital (e.g., Facebook), whose main advantage is to extend the teacher-student relationship outside the traditional school spaces and times. But this mediation requires good communication skills from the teacher, which are not widespread.
b) Microblogging (e.g., Twitter), which enables new forms of exchange and online storage in order to be used as study resources, as if they were a handout. Students and teachers can exchange information and

ask questions, allowing new interactions at the level of the classroom and out of it, but can also save Tweets to use them for reviews and for the teacher to realize the extent to which students understand content.

c) Bookmarker platforms, such as Diigo, which facilitate the maintenance and/or sharing of resources available online, with associated annotations. The bookmarks serve as memory aids and the bookmarked resources are potential study and work documents. The action of bookmarking resources can be a collective effort as these platforms enable users to follow people who do similar work or interest to the user. By creating and sharing a bookmark they have indicated that the resource is of some quality and importance to them.
d) The thematic, such as Ravelry (which is a crochet and knitting OSN), which enable you to share information, tutorials, and explanations that can be used in self-training. Users can share their work for free or counter and return and even point out the type of work they like, so that the relationships created are mainly for exchange and learning.

OSN have educational potential, but the fulfillment of that potential requires an understanding of the relationships between individuals and the media, because only by knowing these relationships is it possible to identify potentialities, capacities, and needs of those individuals in relation to the *media* usage. It is necessary to make a diagnosis to justify the action. The action is to prepare all citizens for the inter-relationship with the media and the whole society through these media.

Citizens have to develop competences at five levels:

a) **Techniques** for dealing with digital technologies, which implies capabilities in terms of *hardware*, *software*, programming, and security monitoring (UNESCO, 2014);
b) **Reflective and creative**, to achieve the critical analysis and production of media messages (Buckingham, 2009; UNESCO, 2011), which implies the ability to evaluate contents, reflect on the political and economic influence in the media, and know how to deal with risks and opportunities (UNESCO, 2014);

c) **Self-regulation (subject-media)**, which is the use of technology and media in a personal logic, exercising rights such as being connected or not, deleting the digital footprint, using media innovatively and creatively whether for entertainment, learning or work (UNESCO, 2014);

d) **Social or general**, crucial to an effective participation in contextualized practices that only make sense together with the above, in a holistic way, since "the skills cannot be understood out of context" (Meyers et al., 2013, p. 361);

e) **Pedagogical**, which consists of knowing how to use the media in learning situations, not just "to know its potential and its risks," but rather "to know and know how to use their languages and codes in the critical perspective of a broad learning that today is an essential part of the learning experience" (Fantin, 2012, pp. 71–72).

The range of skills associated with Media Education can be considered under the broad concept of "transliteracy," which can be defined as "the ability to read, write, and interact across a range of platforms, tools and media, from signing and orality through handwriting, print, TV, radio and film, to digital social networks" (Thomas, 2008, p. 101). We associate "transmedia" with this concept, which depends on the set of competencies of individuals to "decode, remix, create, and circulate many kinds of media content... across contexts, including school, extended learning programs, and home environments" (Alper & Herr-Stephenson, 2013, p. 366).

This preparation should begin in the cradle (Gonnet, 1999) and continue on a lifelong basis (Rivoltella, 2012), in informal and non-formal contexts, but also in formal contexts and in school. This citizens empowerment should be done through Media Education, as has been advocated in UNESCO documents (1982, 2007, 2014), the European Union (Commission of the European Communities, 2001, 2007, 2009; European Commission, 2007, 2008) and governments of various countries in Europe, including Portugal (Ministério da Educação e Ciência, 2011). But there are a number of challenges to which we must respond so that potential might be achieved (Tomé, 2015), namely: to break the traditional school resistance to change, to integrate formal and informal learning, to educate for the Media beyond the technical dimension, to train teachers, to fight the digital divides, and to protect the data of citizens.

Besides creating conditions to meet the challenges, the need to develop field studies in the digital media and learning (DMAL) area is now clear, although not focused on how digital tools can help you learn, but on how these tools that are associated with new forms of convergence, social organization, and complexity of popular culture, "can teach us how to enhance learning in and out of school and how to transform society and the global world as well" (Gee, 2010, p. 14).

OSN are one of these tools, with the advantage of "[allowing] educators to meet their students in their own personal environments" (De Abreu, 2011, p. 100), but this requires that parents and teachers speak their language because if they know that their children and students use OSN, amuse themselves on OSN, and spend a long time there, the truth is that "they haven't found insight into this part of their students' lives" (De Abreu, 2011, p. 52), the practices they develop there, how they communicate, and how they learn on these platforms (Ito et al., 2009).

In this chapter, we want to collaborate on the diagnosis that means that if young students (9–16), their teachers, and parents consider OSN as a way in which they can be prepared for the "new media ecology" (Ito et al., 2009, p. 6), which is becoming increasingly complex. We analyze the perceptions of young people, teachers, and parents in relation to learning in and through the OSN. We approach the relationship among members of the three subsamples and we analyze the opinions of students about a class Facebook page where teachers publish content relating to classes, homework instructions, and other information.

Adults' Positions in Relation to OSN Use by the Young

A third of teachers did not talk to students about OSN, but 65% argued with their attitudes on OSN to be avoided. Three quarters of parents talked to the students about OSN, and a third (32%) have defined usage rules, a third (31%) discussed with the students about what they did on OSN, and another third (38%) explained that there are attitudes to avoid on OSN (Figure 1).

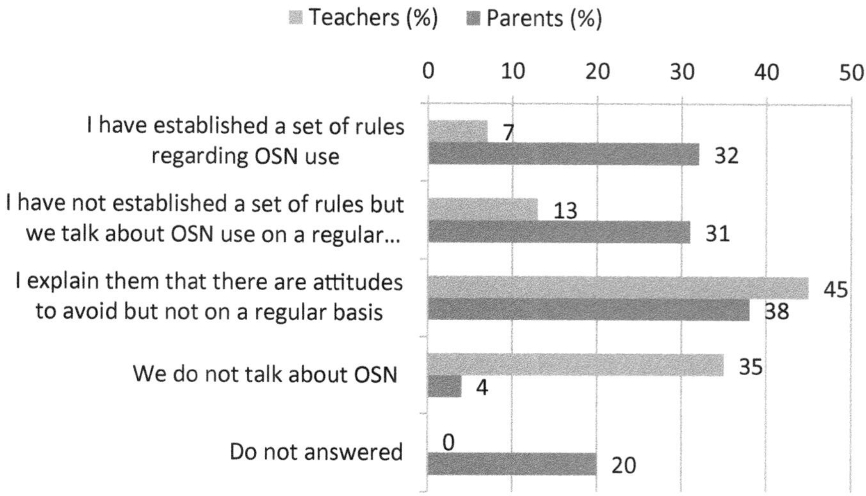

Figure 1. Teachers' and parents' positions regarding the use of OSN by young people (%) N = 150 (teachers); 267 (parents).

We organized teachers into three groups, which were confirmed in interviews during the qualitative phase of the research:

a) A group that warns of the consequences of certain behaviors on OSN, but do so by making unilateral decisions on behalf of the student user --without student participation or student discussion. These actions were stated in situations such as:
   - Non-curricular Education for Citizenship, taught by the class director in which it is mandatory ("When I was a class director, I did it. I am no longer I do not do..."; T22);
   - ICT ("That is one of my fights in the subject. It is very important for them, as they do not often have a sense of danger"; T23), which does not exist in all school grades ("We only have in the seventh and eighth. It was stupid to have finished with ICT we transferred the responsibility to the class directors" T7).
   - English ("We have a unit that is exactly about the dangers of the Net in the unit of New Technologies"; T8);
   - Geography ("There are things that even have to do with the curriculum. When I speak of globalization"; T24).

b) A group that, with or without established rules, discusses with students about OSN and practices on OSN. They are "talks of circumstance, sporadic" (T1), not organized "in a purposeful way" (T10), and sometimes are of the students' initiative ("Sometimes they ask. I speak a lot of incidents that happened, to warn them of what might happen to them"; T24). Students' experiences are used as anchors ("When I talk to them, sometimes they give examples, or own experiences or incidents they hear. Sometimes they accomplish"; T25). Teachers use familiar examples:

- "I gave them the example of a film, *The Trust*. They have seen it, and told them to see the result. I always tell them they do not know who is on the other side. They say everything, what time they leave, what they are wearing, which class are they" (T23).
- The issue of Internet security and the OSN in particular "was discussed in the school newspaper" and "is approached by colleagues whenever ask students to research something on the Internet" (T7).

c) A group that does not speak about OSN with students because they have never thought about ("it never occurred to me to approach that subject"; T13), because there is not enough time ("I have an extensive curriculum and it is not part of it, is not addressed, and that is it"; T2), or because of a lack of information about OSN ("I often do not know what to say because I do not know much about social networks"; T14). These teachers also justify their position by saying that there are specific resources in school for students regarding OSN issues such as "a training session on Internet safety" (T13), the Safer Internet Day (T2), or "the lecture of the School Safe" (T40).

We also organized the parents into three groups, two of which were confirmed in interviews as all parents who agreed to be interviewed, who used OSN, and who spoke with the students about the use of OSN:

a) A group has formal rules and checks if the students comply ("I help him to better use, what to do, what not to do"; P1). The rules concern how long he uses it ("We try to limit such time, which is not easy because the access is made by mobile phone and we do not know whether he is playing or communicating"; P41), and not using OSN alone ("He cannot stay in his room only in the living room"; P16). There are frequent warnings to not reveal personal information, and there are parents interacting with students on OSN ("We share useful information for her on her profile. She likes and sometimes returns, sends interesting information"; P19) because "we need to know them well, to tell them that we trust them. More than forbid, we must tell them, 'I trust you'" (P27).

b) A group with no established rules discusses practices on OSN with the students and gives them advice. Talks take place "occasionally" (P12) or in a more organized way ("We have a family activity once a week and that is when we debate what is new, what is necessary to plan"; P6), which ends in "mutual learning" (P1). Parents follow the example because they are "constantly listening to things and really want to tell them" (P4) and organize "from time to time, scary seminars to tell those awful incidents that happen" (P41).

c) A group that does not speak with the students about OSN (4%) or did not even answer the question (20%), which may be associated with lack of interest or ignorance about OSN.

In summary, there are conversations about OSN among parents and students and among teachers and students. Those conversations take place in the school's formal and non-formal space, in a compulsory and non-compulsory way, and in the informal space at home, as admitted by 96% of parents who answered the question. But they are mainly focused on the dangers that young people will be exposed to when using OSN and not the opportunities:

> There is still a large group of teachers who have a great knowledge of what the social networks are and its use. I think that group, uninformed, does not see anything else on the networks except the dangers. And it should not be only 35% who do not speak on networks with students. It must be many more. (T40).

Parents even limit usage time, because they cannot mediate the use of the devices, especially mobile devices, as the smartphone is a very private device and difficult to monitor, even within a close family environment (Mascheroni & Cuman, 2014). This fear of teachers and parents can also result from a lack of knowledge about how OSN work, which means that many of them do not feel prepared to discuss them with young people, thus limiting the use of OSN for educational purposes.

Educational Potential of OSN: Teachers' Perceptions

Seven out of 10 teachers admitted that social networks have potential for educational use, namely: (1) 78% of teachers who used OSN and 48% of those who did not use them; (2) 96% of teachers who used OSN for educational purposes and 60% of those who did not; and (3) 88% of teachers who spoke with students about OSN (e.g., uses, contents, protection strategies) and 64% who did not do so (Figure 2).

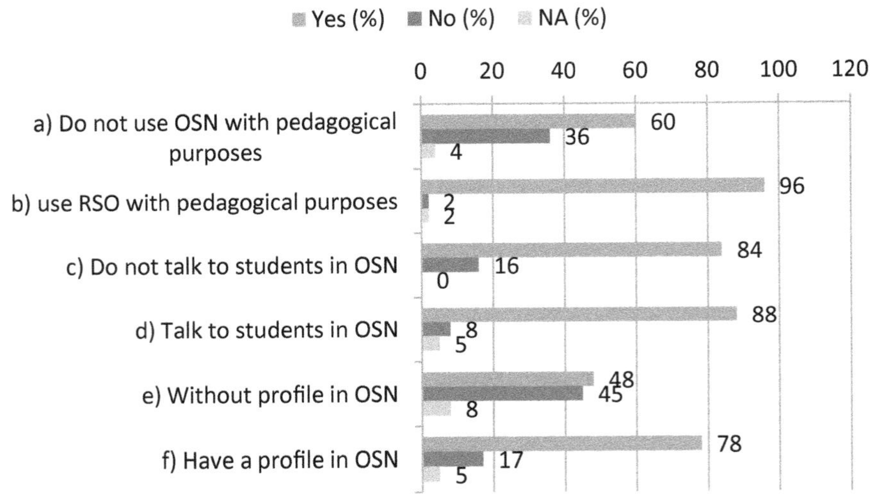

Figure 2. Pedagogical potential evaluation of OSN by teachers (%) N = 100 (s); 47 (b); 85 (c); 18 (d); 40 (e); 110 (f).

The margin of progression in the use of OSN for educational purposes is high, since three out of four teachers have at least one profile on OSN, with about 90% of teachers and students, as well as parents, using Facebook. Such common usage may not necessarily mean contact among teachers, students, and parents, as is shown in the data analysis, but 36% of teachers talked to students on OSN, so there is room to grow that percentage, and the dialogue or asynchronous interaction with educational objectives, especially when 43% of teachers said they already use OSN with these objectives. At this level, there is still a long way to go.

OSN Use With Educational Purposes at School

Only 18% of students and 14% of teachers believed there is learning on, and through, OSN for school contents, and 7% of teachers and 2% of students did not answer the question (Figure 3).

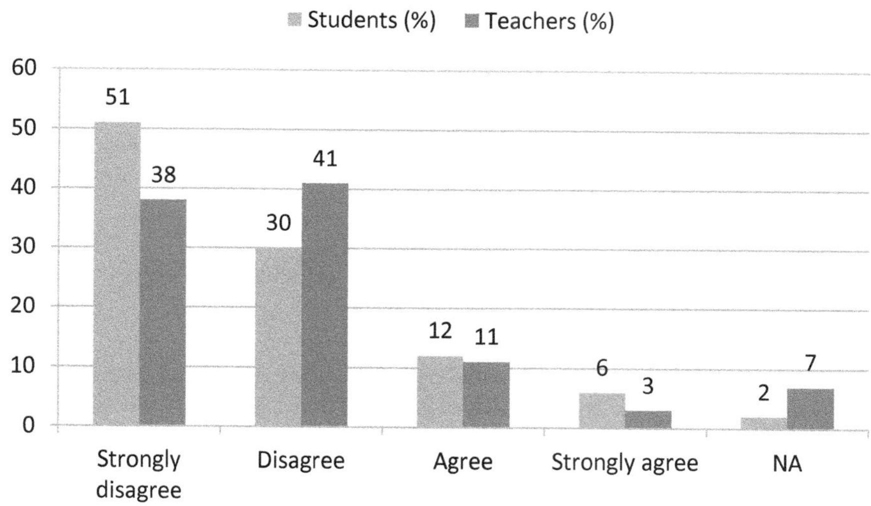

Figure 3. Teachers and students attitude regarding learning school contents on OSN (%) N = 499 (students); 150 (teachers)

The interviewed teachers were divided. A third of the 20 respondents did not admit the pedagogical potential of OSN. "I do not see which pedagogical interest the networks might have" (T2), because "it is more desirable to use the school platforms, since it is much safer and there is no absent-mindedness, such as games" (T22), and "if the network were a study space, research and investigation, they will automatically cease to be interested in that" (T14).

Two thirds of teachers surveyed believed in the pedagogical potential, as "we only have 10 years of networking and the potential will be increasing" as "someday the books disappear and everything is in the tablet" (T40). You have to realize the potential: "They see networks especially as entertainment and communication space. But teachers have the outlook that there is learning" (T25).

Six of the 20 teachers used or had used OSN for teaching purposes. A music teacher had two Facebook pages where she shared music and interacted with students and parents. "I know that this way, I cannot reach everyone. But if two or three learn it is good. I try to use all the techniques to reach the maximum number of students, for them to learn" (T1).

A geography teacher used their personal page to interact with students, sharing YouTube videos. In a test he used an image that he had found on Facebook on a transportation system that students identified immediately. "It was a picture of a bicycle to illustrate transport problems. They said that if they knew, they would have gone to my Facebook" (T24). He still uses OSN for planning:

> In tomorrow's lesson I am going to start the Environment and Society. So I will start by the theme *I'm sorry*, Michael Jackson. You see here, desertification, the monkey moving on to forests, you are seeing water missing....To convey the message you have to have an anchor. (T24)

He used OSN for self-training, interacting with colleagues, with the example of a video about the expansion of the oceans:

> Last week, someone shared that when the oceans freeze they also expand. I automatically shared. I copied it to myself to show in class. So I do not go to Facebook looking for things, but it can happen to find. I am in Coimbra Geography newsgroups and they are always sharing on Facebook. And I take advantage. (T24)

A teacher of Portuguese used YouTube videos. "I went there once to get things on Lusíadas or Fernando Pessoa. I found it interesting. Of course you have to visualize previously. But I thought it was accurate. There was nothing out of the scientific" (T25).

A natural sciences teacher used OSN and other available ICT, but not Facebook:

> I use YouTube to display them videos. The computer in the room is always on, because I have the digital handbook and kids do not. As all the rooms have a video projector, I often want to give examples and use the handbook, links to networks and, although not directly open, they redirect to YouTube. But I never used Facebook with those purposes. (T40)

But it is proven that students use and learn contents that are important in classes:

> In the middle of a lesson sometimes they say: "Somebody put something about it on Facebook." And when we talk about daily news, there are also many that appear on

> Facebook. Then the students ask questions about them, for example when there is an important scientific discovery or when there was a science prize. (T40)

A teacher of arts believed that students develop learning in and through OSN:

> Some time ago I published something about the gray city, São Paulo, which has the gray brigade, who get graffitist and paint everything in gray. Gray has an extraordinary symbolism and kids did a documentary on the subject about the guerrillas between graffitist and the brigade of gray paint. It was very interesting. (T10)

An English teacher used a Facebook page in which he posts "every day several things like vocabulary charts or grammar, sometimes idioms" (P7). He shared with the students, but the shares were received unevenly:

> I sometimes post things there and they say: "Oh teacher, here?" That is, I think it is something unresolved yet. Kids do not still consider networks as work platform, clearly not. They think that it is pure entertainment. If we can get them to go with some injections of something of some work, it has to be a matter that teachers must learn to do....They accept school news, school activities, schedules, because they do not think it is work. While school activities disclosure platform, is not the same. While working platform for the subjects, I think difficult. (T7)

Teachers who believe in the pedagogical potential of OSN were more cautious: "We have no idea what students learn on the OSN. And as we have no idea, we do not take advantage" (T10), assuming that "its pedagogical achievement must be made almost surreptitiously, since only a minority accepts learning school stuff on Facebook" (T7). And consider that "there must be another preparation for teachers and students" (T8).

The students' age influences the perception and learning of school contents on OSN. The students of higher age groups are the ones who tend to value OSN as learning spaces. The percentage of agreement is less than 16% up to 12 years, but rises to 20% at 13 years, reaching 22% at 14 and 35% at 15 years.

A minority of students aged under 12 referred to learning subject contents because "some teachers publish things they might teach us" (Miguel, 12), for example "the teacher of Music publishes on Facebook and also on the website school, and I learned things there that helped me in classes" (Gonçalo, 11). But these students have a hard time giving examples: "I learn a lot of things, but...examples I do not remember any" (Tomás, 11). There were those who considered not learning on OSN, because on those spaces "it is more to spend time and not to learn" (Beatriz, 12). They very often stated that they do not learn anything related to school contents.

A minority of young people between 13 and 15 referred to learning school contents on OSN such as "English" (Inês, 14) and "Geography, because the Geography teacher often shares general culture and things like that" (Beatriz, 14).

School contents come mainly via e-mail to a common account ("Teachers send things to the e-mail"; Francisco, 15) or through the "Moodle, which has things there not just about school" (Bruna, 13). Some of those contents are then shared in private Facebook groups, open to the class, with added benefits, but with limitations, as publications and shares in the group go far beyond the classroom content and bother some students:

> Y3—Instead of sending messages to each, we publish and are to everyone. School information is visible to the whole class, not to others.
> Y7—The problem is that class students use that more to publish news of their clubs...
> Y3—We created more that to talk about class, about school, etc. But I log in and just see, football, football, football.
> Y5—It is true.
> Y3—I already told them to create an account to talk about football.
> INT—But are there school information? Are there stuff of subjects and testing?
> Y5—Yes, there is some.
> Y8—But mostly are ones scolding the others saying who is playing well or who is playing badly.

There are students who resist learning school contents on OSN: "If we want we learn. But that is the problem, we do not want" (Rita, 16 years). But still they learned, as 23% of students said that they agree or totally agree with the item, "I have classmates that, in the social network, help me to better understand the lessons' contents." In the interviews they confirmed: "I sometimes before the Physical Chemistry test, well, I have a site where everything is well explained. I already shared with Pedro and Luis" (Alexandra, 13).

Students showed a preference for learning methodologies associated with project work, which integrate formal and informal context, as is possible through the OSN, in terms of access to content, content creation, connection with other learners and a more personalized relationship, creative and collaborative between teacher and students (Redecker et al., 2010). And they justified this option with the fact that "when an older person is explaining, it seems that you are under pressure, and is more difficult to learn and memorize" (Alexandra, 13), "even because those people use a different language"

(Vanessa, 14). And they considered learning on the OSN, through the OSN, and about OSN.

OSN Learning in Non-Formal/Informal Settings

About two thirds of teachers and 43% of students believed that young people learn about subjects that interest them on OSN , although not directly related to school contents (Figure 4).

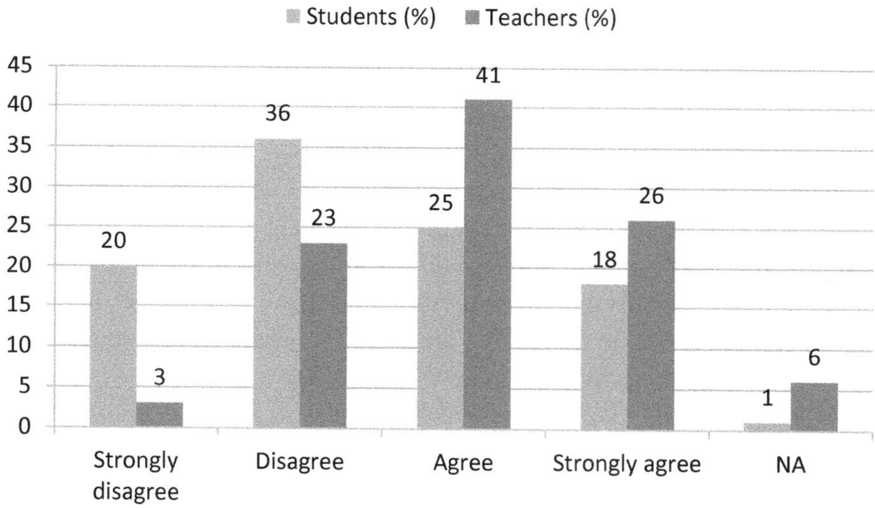

Figure 4. Teachers' and students' opinions regarding learning other content on OSN (%) N = 499 (students); 150 (teachers).

The agreement of children with learning content on OSN, but not the school ones, increased with age, from 22% at 10 years to 35% at 12, 48% at 14, and 56% at 15. Students pointed out contents they learned on the Internet and OSN (they sometimes did not distinguish), which we have organized into seven groups:

a) **Techniques**—learn to "build objects," "edit video," "computing," or to manage the attendance OSN: "we learn that we cannot always be publishing what comes to mind and comment the way we feel like" (António, 12); "We learn to better use Google to search for pages and to know what are the Facebook pages where are things that interest me" (Mariana, 13).

b) **Games**—"I learn about games," "I learned to play snooker on the network, I am a master" (João, 15); "I play a game called League of Legends and there are pages only on this game. They publish a lot. I go there to see and I learn about the game, I learn to play better" (Miguel, 15).
c) **News**—the news was often mentioned, and the news that was of most interest was related to deaths (Eusebio's death, Nelson Mandela's death), sports and stars, whether showbiz (e.g., Justin Bieber) or sport (e.g., Cristiano Ronaldo).
d) **Arts and entertainment**—learn more about music, groups, TV series, and movies. "I put 'like' on pages and then I check, for example, groups, films" (João, 13).
e) **Sensational news stories**—young people often mentioned "anecdotes," "magic tricks," and "curiosities."
f) **Musings**—"important people sentences" to which are not detached: "I see phrases that are life lessons and think about it" (Madalena, 12); "This can be highly stupid, but there are certain moral lessons that help me understand things better. There are some stories about life that are good, no doubt. We are still very young and some sentences and things help to think a bit" (Marta, 13).
g) **English**—the English they learn is for non-academic purposes. "They are phrases that we hear and then we use among us" (Mary, 15); "I learn English on music and in the series" (Inês, 13).

Learn to Use the OSN Through the OSN

Students considered learning to better use the OSN through the OSN, which was more referred to as the age increases: at 10 years only 20% admitted it, at 11 there were already 28%, rising to 35% at 13 and ranging between 44% and 45% at 14 and 15 years.

Students of 10–12 years talked about learning to protect their password ("If my friend had his pass stolen, I know I will be more careful"; Gonçalo, 11), to prevent viruses ("my cousins know a lot about computers and they told me not to press some links, because they are viruses; I learn a lot from them"; Duarte, 11), or to block friends ("I still do not block, but my cousin came over the other day and blocked a colleague here at school"; Inês, 7th grade, 12 years).

They learned from family but not teachers, except "those who come from outside" (Gonçalo, 11) or "with the CD [class director]" (Fábio, 12). With parents, they said that they learned little or nothing: "Sometimes I ask something to my mother and she says: 'Do you think I understand anything of computers to answer you that?'" (Maria, 11).

Students aged 13 and up learned "to better understand the dangers of the network" (Marcelo, 13). They learned with colleagues and family members, but less with parents and more with some teachers, who understood the problems better than their parents did: "Teachers go through many students and learn from them. Our parents only live with us. And when there are news about it soon become very concerned" (Liliana, 16 years).

Learning in the OSN: Parents' Perspectives

According to the data collected in the interviews, 14 of the 20 parents considered that students "learn in networks, but not school issues" (P41). Only six admitted learning school contents on the OSN or through the OSN ("My niece told me that she studied for Physical Chemistry test on Facebook, sharing information with her friends"; P1), whose potential should be further explored:

> When I need to help my daughter, in certain tasks, I do not remember the subjects and I have to do research. I meet very friendly teachers, particularly in Brazil and other countries, in which they make videos explaining the subjects as if they were in class. In Portugal we do not see much. There is no such habit. It is not instilled that tool, which is great and they could use in a willful way. (P17)

Parents also considered that teachers could share tests and test keys on OSN, which "would help to help our children" (P4), because "in Mathematics I even know, but History, I have to study to perceive if he is replying well" (P7). But they said that "such a process has to be coordinated and work well, otherwise our children use the computer just to do what they like, play, and not also learn" (P14). Although they agreed that they were learning in and through OSN, they also pointed out limitations: "You cannot forget the human contact that is essential in learning, but networks can be a path, not the path, but a path" (P19).

OSN Use in Schools: Desired, But Limited

If teachers and parents recognize the educational potential of OSN, the truth is that there are limitations to the use of OSN in schools, following an action taken in March 2014 by The General Directorate of Education and Science, a department of the Portuguese Ministry of Education (Lusa News Agency, 2014). Since then, access to OSN such as Facebook, Instagram, and Tumblr was blocked from 8:30 a.m. to 13.30 p.m., being restricted to a certain number of accesses in the afternoon. Access to YouTube has been limited throughout the day.

The measure was taken "following several computer attacks recorded since the beginning of the year [2014], which led to a deterioration in the quality of Internet access" (Lusa News Agency, 2014, para. 10) it having been found that "the pressure on the network stemmed from access to certain websites and applications that, in general, are not of educational nature" (Lusa News Agency, 2014, para. 11).

Considered to be positive "in some situations" by the National Association of Directors of Public Schools, "since there are students who spend the classes in social networks" (Lusa News Agency, 2014, para. 11), the measure was rejected by the National Association of Information Technology Teachers (NAITT), for whom the Internet access should be free, adding that the measure prevents the teaching of the subject content of information and communication technologies (ICT) about OSN and the caution required when using it.

Not all teachers were aware of the effects of Internet access limitations and OSN: "In relation to the students, do not know. In relation to teachers, there is access, because the school has Facebook. I think they do not forbid" (T1). Some even mentioned free use: "In my school are allowed, from the library to the computer in the classroom" (T10).

The most common situation in schools is the prohibition of mobile phone use and OSN in the classroom, limited access in the library and school area, via computer or mobile phone, since "games are not available, but they have access to the profile" (T24). Students are informed about the constraints and do not access OSN through computers available in the library or in other spaces:

> Y4—Laptop is more important or even the mobile phone, since the school computers sometimes do not give access to sites and sometimes even to Facebook.
> INT—The computer school is locked?

> Y5—You enter the Facebook and suddenly, the page closes. On YouTube happens the same.
> Y6—Some say: "This site is off link by government orders." The Government instructs school to not allow. I think it is to prevent problems in school. I think it is so...that is why they say so.

They managed to thwart these constraints, not through the laptop, which they gradually ceased to take to school, but on their mobile phones, and "there are students who have more powerful mobile phones than teachers" (T10).

> A7—"We can bring our laptops and use the school network. But if we try, for example, to access some websites, it says that is blocked by the Ministry of Education."
> A5—"I go to Facebook at school, but on the mobile phone."
> (Excerpt from the interview with a group of seventh grade students, 12/13)

The frequent use of mobile phones by students was confirmed by the teachers:

> Students are something curious. Last week I saw the responsible for the school computing desperate because all morning, the school data system was completely overload due to Androids. Mobile phones are constantly connected. It is a massive flow of Android access to the school network and creates technical problems. (T10)

We asked students if they would like to have a class Facebook page that classmates, teachers, and eventually parents could join, where teachers could publish school contents, homework, and other information. Most students agreed, but with constraints.

More than two thirds of students (69%) agreed with the creation of a class Facebook page. They considered that creating a class page would enable the use of OSN for useful purposes, to become better known, and for fun. But the most cited advantages were related to how it could be helpful to understand the school contents, or to help with queries and support in testing, as well as support in homework and other work. The class page could also help students from other classes.

The fact that the page would be available on the Internet was considered advantageous because students wanted to "learn using technology," a feature that facilitates learning, and enables them to learn "more and better." It also enables them to "share information" from the school, not only between students and teachers, but also with others, such as parents. Next are the benefits associated with bigger and better "communication and collaboration" with colleagues, with teachers, between students and teachers when carrying out group work, but also among teachers.

Agreement lessens as we progress in the age group, being 81% at 10 years, 65% at 13 years, and 55% at 15 years, but there are always more than half of the students in each group to agree to the creation of the page.

About a third of students (31%, 154 students) did not agree with the creation of a class page on Facebook, arguing that the page "would not work," "would give more trouble" it was unnecessary because "the classes are enough" and "social networking is not to have school contents." In the interviews these reasons were replaced by the need to avoid exposure on OSN at four levels:

a) **To their peers**: As a group, they contended that they should not proceed with the page "or the other classes would snoop what we were doing" (Joaquim, 11).
b) **To their parents**: "No one will want to know what we do in class, except our parents" (Manuel, 14) and if they know "they can mock us" (Fátima, 11).
c) **To their teachers**: To create the page under the proposed conditions, "the teachers would have" access to the students and staff profiles" (Fábia, 12).
d) **To strangers**: Students feared that "many could see or change something" (Mónica, 13), "someone stupid would soil this group" (Pedro, 12), or that "any class member could say things or write there and then blame me" (Margarida, 11). "It could [still] get there a hacker and find out information" (Victor, 15).

Other reasons mentioned by students include the fact that some teachers and students do not have a profile on OSN, do not like to use them, or do not know how to use them, which would exclude them from the class page on Facebook.

In some cases (25 out of 142 interviewed students) the page already exists. In such situations the page consists of a private group created by one or more members of the class, and it ceases by the end of the school year, so the students create another private group the following year. Students referred to sharing in those groups, tests, worksheets, test dates, information on homework, besides clarifying some doubts. They still share photos and other information not related to school content.

Only the students who are on Facebook have access to those groups. Teachers do not have access, as shown by the results, since teacher participation in Facebook students groups has only been referred in two of the focus groups interviews. The wish for the non-participation of teachers is widespread, regardless of age. The younger, under 12, age group argued that "if I ask there what is the homework, the teacher knows straight away that I do not know because I was not paying attention" (Maria, 11) and that "if anyone write or show there a class secret, teachers would know" (Diana, 11). From 13, the arguments are stronger: "It is a matter of privacy" (Elisa, 16); "We do not want teachers there. The class group should be like a group of friends" (Jorge, 15).

Remember that teachers send some information to the e-mail box of the group (e.g., tests' matrices) that they do not participate in, "or, the teacher saw what we sent there." They concluded: "With regard to school, e-mail is more for teachers. Facebook is for us" (Pedro, 16).

Conclusions

Although they are more focused on risks than opportunities, seven out of 10 teachers surveyed recognized the pedagogical potential of OSN, especially those who use them for educational purposes, who have profiles and interact with students through them. And there are examples of good practice in the field and in different subjects, but developed by self-initiative, for "training activities often take place on the sidelines without a really defined program taking into account the needs of teachers who try to integrate, by trial, these devices in their classes" (Le Deuff, 2011, p. 68). The results show that teachers are users of technology and media, but more out of school than in the classroom (Tomé & Domingues, 2014).

The parental respondents also agreed that OSN have educational potential. Some of them used YouTube to find videos and help the students. They thought that young people could learn on OSN, but only contents not related to school. The connection between OSN and the school is thus almost opposite, as only 15% of teachers and 18% of students agreed that there is learning of school subjects on OSN. Learning of non-school content via OSN was accepted by 72% of teachers and 43% of students. English was one example pointed out by students that distinguished the more formal learning at school from the informal learning of OSN, in which they learn with peers, relatives, and others, preferring the latter to the former. On the other hand, teachers considered that they needed support, according to a respondent: "We have no idea what students learn in the OSN. And as we have no idea, we do not take advantage that way" (T10).

If 69% of young people agreed to create a class or group page on Facebook for school and teaching purposes, they made it clear, though, that they did not want teachers on these pages, only admitting the hypothesis of creating two pages, one with teachers' participation and another only for students. A third of young people said that they did not agree with this page as they did not want to be exposed to adults and/or peers, known or unknown.

Most teachers and young people, but also 20 of the parent respondents, separated formal and informal context and did not regard learning from a holistic perspective, where the formal and informal contexts interact and complement each other. The integration of OSN in the formal setting is therefore dependent on a set of changes. It urges change to the school culture, breaking the resistance to change, supporting teachers with training and resources, training whose need they realize themselves, and claim for. It urges the phasing-out of digital divides by promoting fairness. It urges the integration of formal and informal learning contexts, towards lifelong "continuous learning ecologies" (Meyers et al., 2013).

It is necessary to increase the level of digital literacy of all citizens, focusing on teachers, students, and parents. This will only be achieved with the effective integration of Media Education in the curricula, which should cover all media, from printed to digital and not just focused on the latter, as Buckingham (2012) warned: "Do we really need Media Education 2.0? Maybe yes; but, certainly, we also need Media Education 1.0 yet" (p. 54). But we cannot ignore the role that digital media can have in educational terms. It is necessary to continue to do research, develop projects in the field, and evaluate them critically. That is if the educational potential of OSN and other digital technologies are not exploited, "there is a clear field again to the influence of advertising industries, as happened to television" (LeDeuff, 2011, p. 68).

References

Alper, M., & Herr-Stephenson, R. (2013). Transmedia play: Literacy across media. *Journal of Media Literacy Education, 5*(2), 366–369.

Buckingham, D. (2009). Media education policy: The future of media literacy in the digital age: Some challenges for policy and practice. In *Euromeduc—media literacy in Europe*. Brussels, Belgium: Euromeduc.

Buckingham, D. (2012). Precisamos realmente da Educação para os Meios? *Comunicação & Educação, 17*(2), 41–60.

Comissão das Comunidades Europeias. (2009). *Commission recommendation on media literacy in the digital environment for a more competitive audiovisual and content industry and an inclusive knowledge society*.
Retrieved from http://eur-lex.europa.eu/LexUriServ/LexUriServ.do?uri=OJ:L:2009:227:0009:0012:EN:PDF

Commission of the European Communities. (2001). *Comunicação da Comissão ao Conselho e ao Parlamento Europeu Plano de acção eLearning—Pensar o futuro da educação.* Retrieved from http://eur-lex.europa.eu/LexUriServ/LexUriServ.do?uri=CELEX:52001DC0172:PT:HTML

Commission of the European Communities. (2007). Communication from the Commission to the European Parliament, the Council, the European Economic and Social Committee and the Committee of the Regions – A European approach to media literacy in the digital environment. Retrieved from: http://eur-lex.europa.eu/legal-content/EN/TXT/HTML/?uri=CELEX:52007DC0833&from=EN

De Abreu, B. (2011). *Media literacy, social networking and the Web 2.0 environment for the K–12 educator*. New York, NY: Peter Lang.

Delesalle, C., & Marquié, G. (2014). *Twitter: Outil de transformation dans le champ éducatif?, Bulletin d'études et Synthèses de l'Observatoire de la Jeunesse*. Paris, France: Institut national de la jeunesse et de l'éducation populaire.

Endrizzi, L. (2012). Jeunesse 2.0: Les pratiques rélationelles au Coeur des médias sociaux. *Dossier d'actualité Veille et Analyse, 71*, 1–24.

European Commission. (2007). *Report on the results of the public consultation on Media Literacy*. Retrieved from http://europa.eu/rapid/press-release_IP-07-1970_en.htm.

European Commission. (2008). *B-Brussels: Study on criteria to assess media literacy levels—SMART 2008/0005*. Retrieved from http://ted.europa.eu/Exec?DataFlow=ShowPage.dfl&Template=TED/N_one_result_detail_curr.htm&docnumber=89657-2008&docId=89657-2008&StatLang=EN

Fantin, M. (2012). Mídia-educação no currículo e na formação inicial de professores. In M. Fantin & P. Cesare Rivoltella (Eds.), *Cultura Digital e Escola: Pesquisa e formação de professores*. Campinas, Portugal: Papirus Editora.

Gallagher, F. (2011). Foreword. In B. de Abreu (Ed.), *Media literacy, social networking and the Web 2.0 environment for the K–12 educator*. New York, NY: Peter Lang.

Gee, J. (2010). *New digital media and learning as an emerging area and "worked examples" as one way forward*. New York, NY: MIT Press.

Gonnet, J. (1999). *Education et medias*. Paris, France: PUF.

Ito, M., Horst, H., Bittanti, M., boyd, d., Herr-Stephenson, B., Lange,...Robinson, L. (2009). *Living and learning with new media: Summary of findings from the Digital Youth Project*. New York, NY MIT Press.

Jenkins, H. (2009). *Confronting the challenges of participatory culture: Media education for the 21st century*. New York, NY: MIT Press.

Johnson, L., Adams Becker, S., Estrada, V., Freeman, A., Kampylis, P., Vuorikari, R., & Punie, Y. (2014). Horizon Report Europe: 2014 Schools Edition. Luxembourg: Publications Office of the European Union, & Austin, Texas: The New Media Consortium.

LeDeuff, O. (2011). Éducation et réseaux socionumériques: Des environnements qui nécessitent une formation, *Hermès*, 59, 67–73. Retrieved from http://www.cairn.info/article.php?ID_ARTICLE=HERM_059_0067#no12

Lusa News Agency. (2014, March, 26). Internet limitada nas escolas impede alunos de aprender toda a matéria. *I Online*. Retrieved from http://www.ionline.pt/384181

Mascheroni, G., & Cuman, A. (2014). *Net Children Go Mobile: Final report*. Milano, Italy: Educatt.

Meyers, E., Erickson, I., & Small, R. (2013). Digital literacy and informal learning environments: An introduction. *Learning, Media and Technology*, 38(4), 355–367.

Ministério da Educação e Ciência. (2011). *Recomendação n° 6/2011 sobre Educação para a Literacia Mediática*. Retrieved from http://dre.pt/pdf2s/2011/12/250000000/5094250947.pdf

Perulli, E. (2009). Recognising non-formal and informal learning: An open challenge. *Quality of Higher Education*, 6, 94–116.

Redecker, C., Ala-Mukta, K., & Punie, Y. (2010). *Learning 2.0—The impact of social media on learning in Europe*. Luxembourg City, Luxembourg: Office for the Official Publications of the European Communities.

Rivoltella, P. (2007). Realidad y desafíos de la educación en medios en Italia. *Comunicar*, 28, 17–24.

Rivoltella, P. (2012). Retrospectivas e tendências da pesquisa em mídia-educação no contexto internacional. In M. Fantin & P. Cesare Rivoltella (Eds.), *Cultura Digital e Escola: Pesquisa e formação de professores*. Campinas, Portugal: Papirus Editora.

Smith, L., & Clayton, B. (2009). *Recognising non-formal and informal learning: Participant insights and perspectives*. Adelaide, Australia: Australian Government.

Thomas, S. (2008). Transliteracy and new media. In R. Adams, S. Gibson, & S. Füller Arisona (Eds.), *Transdisciplinary digital art: Sound, vision and the new screen* (pp. 101–109). Berlin, Germany: Springer-Verlag Berlin Heidelberg.

Tomé, V. (2015). Redes sociais online: Práticas e percepções de jovens (9–16), seus professores e encarregados de educação. In V. Tomé, E. Bévort, & V. Reia-Baptista (Eds.), *Research on social media: A glocal view*. Lisbon, Portugal: RVJ-Editores, Lda., Science and Technology Portuguese Foundation and CIAC-Algarve University.

Tomé, V., & Domingues, I. (2014). A utilização do computador por alunos e professores do 1° Ciclo dentro e fora da sala de aula: Um estudo de caso em Portugal. In L. Rosado,

G. Ferreira, M. Lemgruber, & E. Bohadana (Eds.), *Educação e Tecnologia: Parcerias 3.0* (pp. 265–300). Rio de Janeiro, Brazil: Universidade Estácio de Sá.

Underwood, C., Parker, L., & Stone, L. (2013). Getting it together: Relational habitus in the emergence of digital literacies. *Learning, Media and Technology*, 38(4), 478–494.

UNESCO. (1982). *Declaração de Grünwald.* Retrieved from http://www.UNESCO.org/education/pdf/MEDIA_E.PDF

UNESCO. (2007). *Paris Agenda or 12 recommendations for media education*. Retrieved from http://www.diplomatie.gouv.fr/fr/IMG/pdf/Parisagendafin_en.pdf

UNESCO. (2011). *Media and information literacy: Curriculum for teachers*. Paris, France: Author.

UNESCO. (2014). *Declaration on Augmented Media and Information Literacy (MIL) in the Digital era.* Paris, France: Author. Retrieved from http://www.unesco.org/new/en/communication-and-information/resources/news-and-in-focus-articles/in-focus-articles/2014/paris-declaration-on-media-and-information-literacy-adopted/

Van Dijck, J. (2013). *The culture of connectivity: A critical history of social media.* Oxford, England: Oxford University Press.

· 8 ·

CONCLUDING THOUGHTS

> Although the specific sites and apps may be constantly changing, the practices that teens engage in as they participate in networked publics remain the same. New technologies and mobile apps change the landscape, but teens' interactions with social media through their phones extend similar practices and activities into geographically unbound settings.
>
> —BOYD, 2014

The concept of this book was to offer the reader an opportunity to consider how mobile technologies filter and change the lives of students, while offering learning potential in and out of the classroom globally. On a daily basis, students and adults are using their mobile phones to connect, coordinate, and communicate. It is a staple in their lives and one that is not going to disappear, but grow and change.

As an educational tool, it is beginning to be used with more frequency, but not necessarily with complete confidence or with good learning strategies. Further, while progress has been made, the cell phone or mobile tool is still looked upon with trepidation by adults—whether it is parents or teachers. The consideration of its value in changing the dynamics of the classroom culture is still being judged on whether it is a positive or negative. Studies have shown that both positions can be experienced and verified whether it is in learning or personal development. Because of these variances, there is a subtle underlying doubt of the value of mobile technologies and the impact in learning. This book emphasizes a more deliberate and positive outlook with an emphasis on digital media literacy as the backbone for that direction.

In the first half, themes and issues conceptually serve to highlight how mobile technologies are changing our educational landscape from the connections to digital media literacy and the idea of creativity in the classroom to understanding the influence of this powerful tool as a means to engaging student audiences all over the world. These are areas that are questioned often as perhaps being lost to students in their developmental stages because the mobile technology tool can be used as a babysitter versus an instructional tool. This point is noted in the book, but the act of engagement can also be occurring, and it is one that needs to be discussed further and developed especially for educators in professional development workshops and other formats in which educators can be given opportunities for experimentation with these tools. To assume that an educator will be ready to implement teaching and learning with a mobile tool is incorrect. Throughout the process of writing this book, several conversations were had with current educators who train novice teachers. There were declarations often made that new teachers are more digitally savvy and therefore were able to implement usage of a mobile tool more readily. When questioned further about literacy and how that was used as the guide, there was a negative reaction. In fact, one educator stated, "isn't that the job of the reading teacher or the library media specialist?" No. Digital literacy or digital media literacy is a necessary part of the dialogue when it comes to instructional development of a mobile tool within all curricular areas. Expanding learning to include digital storytelling and the conceptualization of any given application as it applies to classroom learning is a crucial element of the discourse of mobile learning. Handing a tool to a student to use an application without the foundation or that preface to the actual literacy components limits the capability of the mobile technology to the tool. And, tools keep changing and growing. Our task as educators needs to extend to broadening the transliteracy aspects of this growth and make the applications viable learning options for fostering knowledge that will continue once students are outside of the school.

This same thinking must apply to the educator who sees the tool, but may not always understand how to bridge those extensions to the learners they have in a classroom. Further, just because an educator has grown up with technology does not mean they understand this second tier of learning and application in using the technology in a way that has useful value to the instructional elements of core content and curriculum.

At the same time, the greatest and most evolving point in the book has to do with privacy and student data. At the time of writing this chapter, it was already changing. The importance of this issue can be seen in the way in which many conferences all over the world are devoting a great portion of their offerings to workshops that increase or share knowledge on this broadening topic. Whereby before the discussions of issues related to student data appeared to not receive as much national or international focus, the subject has taken a strong turn in the direction of parental involvement and also a raising of consciousness that makes the subject ongoing, and the research evolving.

The original purpose for the collection of school data was to help with tailoring instruction, but regrettably, in most cases, this has not occurred, as some of the data has been misappropriated by companies who work with schools for other business purposes. More importantly, parents were not aware of the initial collection of data and what exact purpose it would serve. This point has lead to the more pressing issue of how the data is safeguarded. As the National Association of State Boards of Education (NASBE) stated, "Unfortunately, the data that schools use to improve instruction are not always adequately protected, and they are often disconnected, decentralized, or aggregated in a way that leaves the information vulnerable to attack" (Vance, 2016). This continues to be a national and international struggle for schools and communities to adhere to school needs while also being proactive in protecting student information.

The substantial empirical study from Portugal in the second half alludes to many of these points as well as many other aspects that were brought about in the first half of the book. The study in some ways follows the premise of the *EU Kids Online* investigation that has been conducted all over Europe and funded by the EC Safer Internet plus Programme from 2006 to 2009 (*EU Kids Online*, 2014). The researcher focused directly on the online social networks' use by students (9–16), teachers, and parents, as well as on the most common practices they develop online, namely news consumption, communication with others, posting and sharing habits, and online relationships among them. Then, the researcher collected focused on risks and behaviors as viewed by both teachers and parents. In the research, it was evident that the fear of the unknown when it comes to online behaviors overtook any understanding or notion of constructive opportunities for students with regards to social media.

As the researcher indicated, the mainstream media has increased these fears substantially as their focus tends to be on the negative and rarely on any positive aspects of being online, especially as it relates to young adult interactions. Later, when looking at learning perceptions, the researcher focused on the potential learning possibilities available through online social networks while also examining the reasons why online social networks are barely used in the classroom. The most obvious predicament was the lack of knowledge of the possibilities of using applications and the usefulness of social networking. However, in general, the usefulness of various technologies was always in question. More importantly, this body of work provided a segmentation and reiteration of a growing problem with learning development in the context of social networks, but also with mobile learning overall.

At the end of the day, the hope is that whether it is through this study in Portugal, the United States, or elsewhere, that educators around the world—North America, as much as in Europe, Australia, or Asia, take away some thoughts and ideas of the influence of mobile technologies and student interactions. Mobile tools are here to stay and the underpinning of growing knowledge and understanding is best served through digital media literacy as the greater context and overarching inquiry.

References

boyd, d. (2014). *It's complicated the social lives of networked teens*. New Haven, CT: Yale University Press.

EU Kids Online. (2014). *EU Kids Online: findings, methods, recommendations*. Retrieved from http://eprints.lse.ac.uk/60512/

Vance, A. (2016). *Policymaking on education data privacy: Lessons learned. Education Leaders Report*. Alexandria, VA: National Association of State Boards of Education. Retrieved from http://www.nasbe.org/education-leader/policymaking-on-education-data-privacy-lessons-learned/

APPENDICES

APPENDIX A

US-DOE- National Education Plan Conclusions and Recommendations

REPRINTED WITH PERMISSION:
U.S. Department of Education, Office of Educational Technology, Future Ready Learning:
Reimagining the Role of Technology in Education, Washington, D.C., 2016.

These are conclusions and recommendations provided by the National Education Plan from the United States Department of Education.

Section 1: Learning

- States, districts, and post-secondary institutions should develop and implement learning resources that embody the flexibility and power of technology to create equitable and accessible learning ecosystems that make learning possible everywhere and all the time for all students.

- States, districts, and post-secondary institutions should develop and implement learning resources that use technology to embody design principles from the learning sciences.
- States, districts, and post-secondary institutions should take inventory of and align all learning technology resources to intended educational outcomes. Using this inventory, they should document all possible learner pathways to expertise, such as combinations of formal and informal learning, blended learning, and distance learning.
- Education stakeholders should develop a born accessible standard of learning resource design to help educators select and evaluate learning resources for accessibility and equity of learning experience.

Section 2: Teaching

- Provide pre-service and in-service educators with professional learning experiences powered by technology to increase their digital literacy and enable them to create compelling learning activities that improve learning and teaching, assessment, and instructional practices.
- Use technology to provide all learners with online access to effective teaching and better learning opportunities with options in places where they are not otherwise available.
- Develop a teaching force skilled in online and blended instruction.
- Develop a common set of technology competency expectations for university professors and candidates exiting teacher preparation programs for teaching in technologically enabled schools and post-secondary education institutions.

Section 3: Leadership

- Establish clear strategic planning connections among all state, district, university, and school levels and how they relate to and are supported by technology to improve learning.
- Set a vision for the use of technology to enable learning such that leaders bring all stakeholder groups to the table, including students, educators, families, technology professionals, community groups, cultural institutions, and other interested parties.

- Develop funding models and plans for sustainable technology purchases and leverage openly licensed content while paying special attention to eliminating those resources and tasks that can be made obsolete by technology.
- Develop clear communities of practice for education leaders at all levels that act as a hub for setting vision, understanding research, and sharing practices.

Section 4: Assessment

- Revise practices, policies, and regulations to ensure privacy and information protection while enabling a model of assessment that includes ongoing gathering and sharing of data for continuous improvement of learning and teaching.
- States, districts, and others should design, develop, and implement learning dashboards, response systems, and communication pathways that give students, educators, families, and other stakeholders timely and actionable feedback about student learning to improve achievement and instructional practices.
- Create and validate an integrated system for designing and implementing valid, reliable, and cost-effective assessments of complex aspects of 21st century expertise and competencies across academic disciplines.
- Research and development should be conducted that explore how embedded assessment technologies such as simulations, collaboration environments, virtual worlds, games, and cognitive tutors can be used to engage and motivate learners while assessing complex skills.

Section 5: Infrastructure

- Ensure students and educators have broadband access to the Internet and adequate wireless connectivity, with a special focus on equity of access outside of school.
- Ensure that every student and educator has at least one Internet access device and appropriate software and resources for research, communication, multimedia content creation, and collaboration for use in and out of school.

- Support the development and use of openly licensed educational materials to promote innovative and creative opportunities for all learners and accelerate the development and adoption of new open technology–based learning tools and courses.
- Draft sustainability plans for infrastructure concerns that include upgrades of wired and wireless access as well as device refresh plans and sustainable funding sources while ensuring the safety and protection of student data.
- Create a comprehensive map and database of connectivity, device access, use of openly licensed educational resources, and their uses across the country.

APPENDIX B

Student Privacy Pledge Signers—Here is the list of over 200 companies who have signed the pledge.

3C Institute 3rdquote 5 Star Students Accelify Actively Learn Advanced Processing & Imaging Aegis Identity Aequitas Solutions Aeries Akindi All Covered (Konica Minolta) ALL In Learning Alma	Alpine Achievement Amplify Amtel Apple ASHR AssertID AT&T Atlas Learning Atomic Learning Authentica Solutions AvePoint Benchmark One, Inc. besimpler Blackbaud Blackboard	BloomBoard Bloxx Boardingware BoomWriter BrainPOP BrainQuake Branching Minds Callido Learning Cambium Learning Canvas Capit Learning CaseNEX Castle Learning Chalkup Classcraft

ClassDojo
Classhopper
ClassK12
classkick
ClassLink
Classroom Rocket
Classworks
Clay Piggy
Clever
Code.org
CodeMonkey
College Board
Common Threads
Compass Learning
Connected Data
Continuum Education
Copilot Networks
CPSI
Curriculum Associates
Didax
Drawp for School
DreamBox Learning
Edaura
Edgenuity Inc.
Edmentum
Edmodo
eDoctrina
EDpuzzle
Edsby
Edthena
Education Framework
Edusight
emAPP
Epiphany Learning
eScholar
ESP Solutions Group
eSpark
Essential Education
Evan-Moor
Examity
FastBridge Learning
Finalsite
Flocabulary
FocalPointK12
Follett
FreshGrade
FreshSchools
Fuse School
Gaggle
Genzer Consulting
GetEdu, Inc
GlassLab
GoGuardian
Google
Gradeable
Gradelicious
Haiku Learning
Hazara
Healthmaster
Homework Friendzy
Houghton Mifflin
Harcourt
HR Imaging Partners
Humble Daisy
IDconnect
Ideaphora
Ident-A-Kid
identiMetrics
Identity Automation
Illuminate Education
Imagine Learning
Infinite Campus
InfoSnap
Interactive
Achievement
InteractWith
Istation
IXL
JAMF Software
Java School
KEYS 2ENGAGE,
LLC
Khan Academy
Kidaptive
kidhoo
Kids Discover
Kinvolved
Knewton
Knovation
Knowledgehook
Known
Kodable
Language Testing International
Learn Jelly
Learni
LearnKey
Learnmetrics
Learnosity
LearnSprout
LearnZillion
Let's Go Learn
Level Data
Lifetouch
Lightspeed
Listen Current
Longleaf
Lumen Touch
MackinVIA
Maestro
Makkajai
Mandarin
Mangahigh
MARi

MasteryConnect	Proctorio	Student Lap Tracker
Math Games	ProctorU	Student Vox
Matific	Promethean	Summit Public Schools
Measuring Education	PSNI	SunGard
Mentoring Minds	Questi	SureScore
Microsoft	Ravenna	Synrevoice
Mimio	Readorium	Tales2go
MIND Research Institute	Redbird Advanced Learning	TCI
MMS	Registration Gateway	TE21
MobyMax	Renaissance Learning	TeachBoost
My eTutor	RevGen Partners	Teacher Created Materials
MyON	Ripple Effects	Teachley
Naiku	Sangha	Team(You)
National Student Clearinghouse	Scholastico	TenMarks
Nearpod	School Bytes	Texthelp
NedERA	Schoolie	Think Through Math
Nervanix	SchoolInfoApp	ThinkCERCA
Newsela	SchoolMessenger	Thought Cycle
NobleHour	Schoology	Tools4ever
NoRedInk	Schoolrunner	TreeRing
Omnito	Schoolzilla	Trovvit
OneLogin	ScootPad	Turning Technologies
OpenEd	Securly	Turnitin
orgLib	Seesaw	Typing Agent
Otus	Shutterfly	TypingClub
Parchment	Silverback	Unicon
Pear Deck	Skoolbo	Verificient
Peerceptiv	Skooli	VoiceThread
Personalized Learning Games	Skyward	Waggle
PlayPosit	SMART	Walsworth
PowerIEP	smoothwall	William H. Sadlier, Inc.
PrepFactory	SoftChalk	WootMath
PrepToon	Splendid Portraits	WriteLab
PRIVO	Spotlight	XtraMath
	Stackup	Zearn
	STOPit	

APPENDIX C

Student Data Privacy, Accessibility, and Transparency Act (MODEL LEGISLATION)

(Reprinted with Permission: Foundation for Excellence in Education, April 2015)

Summary

The Student Data Privacy, Accessibility, and Transparency Act (the "Act") requires the [State Board of Education/State Department of Education] to make publicly available on a continuing basis an inventory and index of the data elements with definitions of individual student data fields in the statewide longitudinal data system. The Act requires the [State Board of Education/ State Department of Education] to create a data security plan to complement federal and state data privacy laws and policies governing the state longitudinal data system, and to provide student data privacy and security-related guidance to local education agencies. The Act grants authority to the [State Board of Education/State Department of Education] to establish policies for the protection of student data not only in the state's longitudinal data system, but also in data systems maintained by local educational agencies in the state. The Act requires certain third parties that receive student-generated content as a result of K-12 school purposes to implement certain security practices, and prohibit them from engaging in targeted advertising, selling student data, building profiles on students, or sharing student data except under limited circumstances. The Act creates a Chief Privacy Officer within the State Department of Education, whose primary responsibility would consist of ensuring department-wide compliance with all student data privacy and security laws and regulations. The Act also authorizes the Chief Privacy Officer to issue student data privacy and security policies applicable to local educational agencies and to handle appeals by parents regarding their rights under this Act. Finally, the Act adds new annual security and privacy reporting requirements to the Governor and Legislature.

Model Legislation

Section 1. {Title}

(A) This law shall be known and may be cited as the "Student Data Privacy, Accessibility and Transparency Act."

Section 2. {Legislative Intent}[i]

(A) The Legislature acknowledges that student data is a vital resource for parents, teachers, and school staff, and it is the intent of the Legislature to ensure that student data is safeguarded and that students' and parents' privacy is honored, respected, and protected. Student data allows parents and students to make more informed choices about educational programs and to better gauge a student's educational progress and needs. Teachers and school staff utilize student data in planning responsive education programs and services, scheduling students into appropriate classes and completing reports for educational agencies. Student information is critical in helping educators assist students in successfully graduating from high school and being ready to enter the workforce or postsecondary education. In emergencies, certain information should be readily available to school officials and emergency personnel to assist students and their families. A limited amount of this information makes up a student's permanent record or transcript. The Legislature firmly believes that student information is important for educational purposes, and it is also critically important to ensure that student information is protected, safeguarded, kept private, and used only by appropriate educational authorities to serve the best interests of the student. To that end, this law will help ensure that student information is protected and expectations of privacy are upheld.

Section 3. {Definitions}

(A) As used in this act:
(1) "Board" means the State Board of Education;
(2) "Department" means the State Department of Education;
(3) "Student data" means information that is collected and maintained at the individual student level in this state, including but not limited to:
(a) data descriptive of a student in any media or format, including but not limited to:
(i) the student's first and last name;
(ii) the name of the student's parent or other family members;
(iii) the physical address, email address, phone number, or other information that allows physical or online contact of the student or student's family;
(iv) a student's personal identifier, such as the student number, when used for identification purposes;
(v) other indirect identifiers, such as the student's date of birth, place of birth, and mother's maiden name;
(vi) state-, district-, school-, or teacher-administered assessment results, including participation information;
(vii) course transcript information including but not limited to courses taken and completed, course grades and grade point average, credits earned, degree, diploma, credential attainment, or other school exit information;
(viii) attendance and mobility information between and within [STATE] local education agencies;
(ix) the student's gender, race, and ethnicity;
(x) program participation information required by state or federal law;
(xi) disability status;
(xii) socioeconomic information;
(xiii) food purchases; or
(xiv) emails, text messages, documents, search activity, photos, voice recordings, and geolocation information.
(b) such information that:

(i) is created or provided by a student, or the student's parent or legal guardian, to an employee or agent of the school, local education agency, or state education agency or to an operator in the course of the student's, parent's, or legal guardian's use of the operator's site, service, or application for K-12 school purposes;

(ii) is created or provided by an employee or agent of the school or local education agency, including to an operator in the course of their use of the operator's site, service, or application for K-12 school purposes; or

(iii) is gathered by an operator through the operation an operator's site, service, or application for K-12 school purposes.

(4) "Education record" means an education record as definedii in the Family Educational Rights and Privacy Act (FERPA) and its implementing regulations, 20 U.S.C. § 1232g; 34 CFR Part 99.3. An education record does not include the types of student data excepted in FERPA, does not include student data collected by an operator when it is used for internal operations purposes, does not include student data that is not formatted for or expected to be accessed by school or local education agency employees, nor does it include student data that a local education agency determines cannot reasonably be made available to the parent or eligible student;

(5) "Eligible student" means a student who has reached 18 years of age or is attending an institution of postsecondary education;

(6) "Student personally identifiable data" or "student personally identifiable information" or "personally identifiable information" means student data as defined in paragraph (3) that, alone or in combination, is linked to a specific student that would allow a reasonable person, who does not have personal knowledge of the relevant circumstances, to identify the student;

(7) "Aggregate student data" means data that is not personally identifiable and that is collected and/or reported at the group, cohort, or institutional level;

(8) "Provisional student data" means new student data proposed for inclusion in the state student data system;

(9) "Redacted data" or "de-identified data" means a student dataset that is not personally identifiable information because the educational agency or institution or other party has made a reasonable determination that a student's identity is not personally identifiable, whether through single or multiple releases, and taking into account other reasonably available information;

(10) "State-assigned student identifier " means the unique student identifier assigned by the state to each student that shall not be or include the Social Security numberiii of a student in whole or in part;

(11) "State data system" means the State Department of Education statewide longitudinal data system;

(12) "K-12 school purposes" means purposes that take place at the direction of the K-12 school, teacher, or local education agency or aid in the administration of school activities, including, but not limited to, instruction in the classroom or at home, administrative activities, preparing for postsecondary education or employment opportunities, and collaboration between students, school personnel, or parents, or are for the use and benefit of the school;

(13) "Operator" means any entity other than the department, local education agency, or school to the extent that the entity:

(a) Operates an Internet website, online service, online application, or mobile application with actual knowledge that the website, service, or application is used for K-12 school purposes and was designed and marketed for K-12 school purposes to the extent that it is operating in that capacity; and

(b) Collects, maintains, or uses student personally identifiable information in a digital or electronic format; and

(14) "Targeted advertising" means presenting advertisements to a student where the advertisement is selected based on information obtained or inferred from that student's online behavior, usage of applications, or student data. 'Targeted advertising' does not include advertising to a student at an online location based upon that student's current visit to that location or single search query without collection and retention of a student's online activities over time.

Section 4. {Chief Privacy Officer}

(A) The Superintendent shall appoint a Chief Privacy Officer, who shall report directly to, and be under the general supervision of, the Superintendent, to assume primary responsibility for data privacy and security policy, including:

(1) establishing department-wide policies necessary to assure that the use of technologies sustain, enhance, and do not erode, privacy protections relating to the use, collection, and disclosure of student data;

(2) ensuring that student data contained in the State Department of Education student data system is handled in full compliance with this Act, FERPA, and other state and federal data privacy and security laws;

(3) evaluating legislative and regulatory proposals involving collection, use, and disclosure of student data by the Department;

(4) conducting a privacy impact assessment on proposed legislative proposals, regulations, and program initiatives of the Department, including the type of personal information collected and the number of students affected;

(5) coordinating with the Office of the General Counsel and other legal entities as necessary to ensure that state programs, policies, and procedures involving civil rights, civil liberties, and privacy considerations are addressed in an integrated and comprehensive manner;

(6) preparing a report to the Legislatureiv on an annual basis on activities of the Department that affect privacy, including complaints of privacy violations, internal controls, and other matters;

(7) working with the Department Chief Information Officer, General Counsel, and other officials in engaging with stakeholders about the quality, usefulness, openness, and privacy of data;

(8) establishing and operating a Department-wide Privacy Incident Response Program to ensure that incidents are properly reported, investigated, and mitigated, as appropriate;

(9) establishing and operating a process for parents to file complaints of privacy violations or inability to access their child's education records against the responsible local educational agency under Section 8 of this Act; and

(10) providing training, guidance, technical assistance, and outreach to build a culture of privacy protection, data security, and data practice transparency to students, parents, and the public among all state and local governmental education entities that collect, maintain, use, or share student data;

(B) The Chief Privacy Officer may investigate issues of compliance with this Act and with other state data privacy and security laws by the state educational agency and local educational agencies in the state and may:

(1) have access to all records, reports, audits, reviews, documents, papers, recommendations, and other materials available to the Department that relate to programs and operations with respect to the responsibilities of the Chief Privacy Officer under this section;

(2) make such investigations and reports relating to the administration of the programs and operations of the Department as are necessary or desirable; and

(3) in matters relating to compliance with federal laws, refer the matter to the appropriate federal agency and cooperate with any investigations by such federal agency;

Section 5. {Data Inventory and Security}

(A) The [State Board of Education/State Department of Education] shall:

(1) Create, publish, and make publicly available a data inventory and dictionary or index of data elements with definitions of student personally identifiable data fields in the state student data system to include, but not be limited to:

(a) any student personally identifiable data required to be reported by state and federal education mandates;

(b) any student personally identifiable data which is included or has been proposed for inclusion in the student data system with a statement regarding the purpose or reason for the proposed collection; and

(c) any student data that the State Department of Education collects or maintains with no current identified purpose;

(2) Develop, publish, and make publicly available policies and procedures for the state data system to comply with this Act and other applicable state and federal data privacy and security laws,

including the Federal Family Educational Rights and Privacy Act (FERPA). Such policies and procedures shall include, at a minimum:

(a) Restrictions on granting access to student data in the statewide longitudinal data system, except to the following:

(1) students and their parents, as provided by the collecting local education agency;

(2) the authorized administrators, teachers, and other school personnel of local education agencies, and the contractors or other authorized entities working on their behalf, that enroll students who are the subject of the data and who require such access to perform their assigned duties;

(3) the authorized staff of the State Department of Education, and the contractors or other authorized entities working on behalf of the Department, who require such access to perform their assigned duties as authorized by law or defined by interagency or other data-sharing agreements; and

(4) the authorized staff of other state agencies in the State of [State], including contractors or other authorized entities working on behalf of a state agency that require such access to perform their duties pursuant to an interagency agreement or other data sharing agreement;

(b) Prohibitions against publishing student data other than aggregate data or de-identified data in public reports; and

(c) Consistent with applicable law, criteria for the approval of research and data requests from state and local agencies, the State Legislature, those conducting research including on behalf of the Department, and the public that involve access to student personally identifiable information.

(3) Unless otherwise provided by law or approved by the [State Board of Education/State Department of Education], not transfer student personally identifiable data to any federal, state or local agency or other non-governmental organization, except for disclosures incident to the following actions:

(a) a student transferring to another school/local education agency in state or out of state or a school/local education agency seeking help with locating a transferred student;

(b) a student enrolls in an institution of higher education or training program;
(c) a student registering for or taking a state, national, or multistate assessment where such data is required to administer the assessment;
(d) a student voluntarily participating in a program for which such a data transfer is a condition or requirement of participation;
(e) the federal government requires the transfer of student data for a student classified as a "migrant" for related federal program purposes;
(f) a federal agency requires the student personally identifiable data to perform an audit, compliance review, or complaint investigation; or
(g) if the eligible student, student's parent, or legal guardian request such transfer.

(4) Develop a detailed data security plan for the state data system that includes:
(a) guidelines for authorizing access to the state data system and to student personally identifiable data including guidelines for authentication of authorized access;
(b) privacy and security audits;
(c) plans for responding to security breaches, including notifications, remediations, and related procedures;
(d) data retention and disposal policies;
(e) data security training and policies including technical, physical, and administrative safeguards;
(f) standards regarding the minimum number of students or information that must be included in a data set in order for the data to be considered aggregated and, therefore, not student personally identifiable data subject to requirements in this law and in other federal and state data privacy laws;
(g) a process for evaluating and updating as necessary the data security plan, at least on an annual basis, in order to identify and address any risks to the security of student personally identifiable data; and

(h) guidance for local education agencies to implement effective security practices that are consistent with those of the state data system;

(5) Ensure routine and ongoing compliance by the State Department of Education with FERPA, other relevant privacy laws and policies, and the privacy and security policies and procedures developed under the authority of this act, including the performance of compliance audits;

(6) Notify the Governor and the Legislature annually of the following matters relating to the state student data system:

(a) new student personally identifiable data proposed for inclusion in the state data system:

(1) any new provisional student data collection proposed by the Department shall become a provisional requirement to allow local education agencies and their local data system vendors the opportunity to meet the new requirement; and

(2) the [State Board of Education/Department of Education] must announce any new provisional student personally identifiable data collection to the general public for a review and comment period of at least 60 days;

(b) changes to existing student personally identifiable data collections required for any reason, including changes to federal reporting requirements made by the U.S. Department of Education;

(c) a list of any special approvals granted by the [State Board of Education/State Department of Education] under (A)(3) of this section in the past year regarding the release of student personally identifiable data, but such list need not include data shared pursuant to any of the exceptions listed in subsections (A)(3)(a) through (A)(3)(g) of this section;

(d) the results of any and all privacy compliance and security audits completed in the past year. Notifications regarding privacy compliance and security audits shall not include any information that would itself pose a security threat to the state or local student information systems or to the secure transmission of data between state and local systems by exposing vulnerabilities; and

(7) Develop policies and procedures to ensure the provision of at least annual notifications to eligible students and parents regarding student privacy rights under federal and state law.

Section 6. {Student Data Collection and Reporting Restrictions}

(A) Unless required by state or federal law or in cases of health or safety emergencies, local education agencies shall not report to the state the following individual student data or student information:

(1) juvenile delinquency records;

(2) criminal records;

(3) medical and health records; and

(4) student biometric information, unless a local education agency determines the information is necessary for educational purposes, such as for identification for online state assessment, provided the information is deleted when no longer needed for this purpose and not maintained as part of a student's permanent education record.

(B) Unless required by state or federal law or in cases of health or safety emergencies, schools shall not collect the following data on students or their families:

(1) political affiliation;

(2) voting history;

(3) income, except as required by law or where a local education agency determines income information is required to apply for, administer, research, or evaluate programs to assist students from low income families; or

(4) religious affiliation or beliefs.

Section 7. {Restrictions on Operators Use of Student Data}[v]

(A) An operator shall not knowingly engage in any of the following activities with respect to such operator's site, service, or application without explicit written or electronic consent from the student's parent or guardian, or an eligible student:

(1) (a) Use student data to engage in targeted advertising on the operator's site, service, or application, or (b) target advertising on any other site, service, or application when the targeting of the advertising is based upon any student data and state assigned student identifiers or other persistent unique identifiers that the

operator has acquired because of the use of such operator's site, service, or application.

(2) Use information, including state-assigned student identifiers or other persistent unique identifiers, created or gathered by the operator's site, service, or application, to amass a profile about a student except in furtherance of K-12 school purposes. For purposes of this paragraph, 'amass a profile' does not include collection and retention of account records or information that remains under the control of the student, parent, or local education agency.

(3) Sell a student's data. This prohibition does not apply to the purchase, merger, or other type of acquisition of an operator by another entity, provided that the operator or successor entity continues to be subject to the provisions of this section with respect to previously acquired student data that is subject to this Act.

(4) Disclose student personally identifiable data without explicit written or electronic consent from a student over the age of 13 or a student's parent or guardian, given in response to clear and conspicuous notice of the activity, unless the disclosure is made:
   (a) In furtherance of the K-12 school purposes of the site, service, or application; provided, however, that the recipient of the student data disclosed:
      (i) Shall not further disclose the student data unless done to allow or improve operability and functionality within that student's classroom or school; and
      (ii) Is legally required to comply with the requirements of this Act;
   (b) To ensure legal or regulatory compliance or protect against liability;
   (c) To respond to or participate in judicial process;
   (d) To protect the security or integrity of the entity's website, service, or application
   (e) To protect the safety of users or others or security of the site; or
   (f) To a service provider, provided that the operator contractually (i) prohibits the service provider from using any student data for any purpose other than providing the contracted service to, or on behalf of, the operator, (ii) requires such service

provider to impose the same restrictions as in this subsection on its own service providers, and (iii) requires the service provider to implement and maintain reasonable security procedures and practices as provided in subsection (B).

(B) An operator shall:

(1) Implement and maintain reasonable security procedures and practices appropriate to the nature of the student data to protect that information from unauthorized access, destruction, use, modification, or disclosure.

(2) Delete a student's data within a reasonable timeframe not to exceed 45 days if the school or local education agency requests deletion of data under the control of the school or local education agency.

(C) Notwithstanding paragraph (4) of subsection (A), an operator may disclose student data, as long as paragraphs (1) to (3), inclusive, of subsection (A) are not violated, under the following circumstances:

(1) If another provision of federal or state law requires the operator to disclose the student data, and the operator complies with applicable requirements of federal and state law in protecting and disclosing that information.

(2) For legitimate research purposes: (a) as required by state or federal law and subject to the restrictions under applicable state and federal law or (b) as allowed by state or federal law and under the direction of a school, local education agency, or state department of education, subject to compliance with subsection A.

(3) To a state agency, local education agency, or school, for K-12 school purposes, as permitted by state or federal law.

(D) Nothing in this section prohibits an operator from using student data, including student personally identifiable data, as follows:

(1) For maintaining, delivering, developing, supporting, evaluating, improving, or diagnosing the operator's site, service, or application;

(2) Within other sites, services, or applications owned by the operator, and intended for the school or student use, to evaluate and improve educational products or services intended for the school or student use;

(3) For adaptive learning or customized student learning purposes;

(4) For recommendation engines to recommend additional content or services to students within a school service's site, service, or application without the response being determined in whole or in part by payment or other consideration from a third party;
(5) To respond to a student's request for information or for feedback without the information or response being determined in whole or in part by payment or other consideration from a third party; or
(6) To ensure legal or regulatory compliance or to retain such data for these purposes.

(E) Nothing in this section prohibits an operator from using or sharing aggregate data or de-identified data as follows:
(1) For the for the development and improvement of the operator's site, service, or application or other educational sites, services, or applications; or
(2) To demonstrate the effectiveness of the operator's products or services, including in their marketing.

(F) This section shall not be construed to limit the authority of a law enforcement agency to obtain any content or student data from an operator as authorized by law or pursuant to an order of a court of competent jurisdiction.

(G) This section does not apply to general audience Internet websites, general audience online services, general audience online applications, or general audience mobile applications even if login credentials created for an operator's site, service, or application may be used to access those general audience sites, services, or applications.

(H) This section shall not be construed to limit Internet service providers from providing Internet connectivity to schools or students and their families.

(I) This section shall not be construed to prohibit an operator from marketing educational products directly to parents so long as the marketing did not result from the use of student data obtained without parental consent by the operator through the provision of services covered under this section.

(J) This section shall not be construed to impose a duty upon a provider of an electronic store, gateway, marketplace, or other means of purchasing or downloading software or applications to review or enforce compliance of this section on those applications or software.

(K) This section shall not be construed to impose a duty upon a provider of an interactive computer service, as defined in Section 230 of Title 47 of the United States Code, to review or enforce compliance with this section by third-party content providers.
(L) This section shall not be construed to impede the ability of a student or parent or guardian to download, transfer, or otherwise save or maintain their own student data or documents.
(M) Nothing in this section or this Act prevents a state or local education agency and their employees from recommending, directly or via a product or service, any educational materials, online content, services, or other products to any student or their family if the state or local education agency determines that such products will benefit the student and does not receive compensation for developing, enabling, or communicating such recommendations.

Section 8. {Parental request for information; complaints of violation}

(A) Parents have the right to inspect and review their child's education record maintained by the school or local education agency.
(B) Parents have the right to request from the school or local education agency student data included in their child's education record, including student data maintained by an operator, except when the local education agency determines that the requested data maintained by the operator cannot reasonably be made available to the parent.
(C) Local education agencies shall provide parents or guardians with an electronic copy of their child's education record upon request, unless the local education agency does not maintain a record in electronic format and reproducing the record in an electronic format would be unduly burdensome.
(D) A parent shall have the right to request corrections to inaccurate education records maintained by a school or local board of education. After receiving a request demonstrating any such inaccuracy, the school or local education agency that maintains the data shall correct the inaccuracy and confirm such correction to the parent within a reasonable amount of time.
(E) The rights granted to parents in (A) through (D) of this section shall extend to eligible students seeking to access their own education records.

(F) The State Department of Education shall develop model policies for local education agencies that:
  (1) support local education agencies in fulfilling their responsibility to annually notify parents of their right to request student information;
  (2) assist local education agencies with ensuring security when providing student data to parents;
  (3) provide guidance and best practices to local education agencies in order to ensure that local education agencies provide student data only to authorized individuals;
  (4) support local education agencies in their responsibility to produce education records and student data included in such education records to parents and eligible students, ideally within threevi business days of the request; and
  (5) assist schools and local education agencies with implementing technologies and programs that allow parents to view online, download, and transmit data specific to their child's education record.

(G) The Chief Privacy Officer shall develop policies and procedures for parents or eligible students to file a complaint with the state or local educational agency regarding a possible violation of rights under this Act or under other federal or state student data privacy and security laws.
  (1) Parents or eligible students not satisfied with the state or local agency's resolution of the matter may file an appeal with the Chief Privacy Officer.
  (2) The Chief Privacy Officer shall establish a process for receiving and responding to such appeals pursuant to Section 4(A)(9).
  (3) The Chief Privacy Officer, in response to an appeal:
    (a) may dismiss a complaint before taking any other action if the complaint fails to allege any violation of this Act [referring to the entire Act];
    (b) may investigate the allegations in the complaint pursuant to the investigatory authority granted by Section 4(B) of this Act;
    (c) shall, if the complaint is not dismissed, issue a written advisory opinion within 30 calendar days, unless extraordinary circumstances justify an extension of time, after the complaint

is filed concerning whether or not a violation of the parent's or student's rights occurred, which shall be available to the public except for those portions which could reveal the identity of a student or a parent; and

(d) shall refer any possible violations of federal law to the appropriate federal agency or agencies for further investigation.

(H) Nothing in this section authorizes any additional cause of action beyond the process described in this section or as otherwise authorized by state law.

Section 9. {Rules}

(A) The [State Board of Education/State Department of Education] may adopt rules necessary to implement the provisions of this Act.

(B) Upon the effective date of this Act, any existing collection of student data by the State Department of Education shall not be considered a new student data collection in accordance with subsection 3(A)(6) of this Act.

Explanatory Notes

i Language adapted from Idaho's "Student Data Accessibility, Transparency and Accountability Act of 2014," available at http://www.legislature.idaho.gov/legislation/2014/S1372E1.pdf.

ii FERPA defines "education records" broadly as those records that contain information directly related to a student and which are maintained by an educational agency or institution or by a party acting for the agency or institution. (*See* http://www2.ed.gov/policy/gen/guid/fpco/ferpa/students.html.)

Education records.

(a) The term means those records that are:

(1) Directly related to a student; and

(2) Maintained by an educational agency or institution or by a party acting for the agency or institution.

b) The term does not include:

(1) Records that are kept in the sole possession of the maker, are used only as a personal memory aid, and are not accessible or revealed to any other person except a temporary substitute for the maker of the record.

(2) Records of the law enforcement unit of an educational agency or institution, subject to the provisions of §99.8.

(3)

(i) Records relating to an individual who is employed by an educational agency or institution, that:

(A) Are made and maintained in the normal course of business;

(B) Relate exclusively to the individual in that individual's capacity as an employee; and

(C) Are not available for use for any other purpose.

(ii) Records relating to an individual in attendance at the agency or institution who is employed as a result of his or her status as a student are education records and not excepted under paragraph (b)(3)(i) of this definition.

(4) Records on a student who is 18 years of age or older, or is attending an institution of postsecondary education, that are:

(i) Made or maintained by a physician, psychiatrist, psychologist, or other recognized professional or paraprofessional acting in his or her professional capacity or assisting in a paraprofessional capacity;

(ii) Made, maintained, or used only in connection with treatment of the student; and

(iii) Disclosed only to individuals providing the treatment. For the purpose of this definition, "treatment" does not include remedial educational activities or activities that are part of the program of instruction at the agency or institution; and

(5) Records created or received by an educational agency or institution after an individual is no longer a student in attendance and that are not directly related to the individual's attendance as a student.

(6) Grades on peer-graded papers before they are collected and recorded by a teacher. (Authority: 20 U.S.C. 1232g(a)(4))

Section 10. {Effective Date}

This Act shall become effective July 1, [20XX]. However, to the extent any provision of this Act conflicts with a term of a contract in effect before July 1, [20XX], such provision shall not apply to the state agency, local education agency, or the operator subject to such agreement until the expiration, amendment, or renewal of that agreement.

iii Some states' workforce data systems use Social Security numbers for unemployment purposes. Policymakers in these states should consider how to link their states' workforce and education data while also maintaining the highest standards of privacy and security.

iv The chief privacy officer could submit the report to the legislature directly, the report could come from the state superintendent, or it could be a joint submission, depending on the state's governance structure.

v These provisions were adapted from California's Assembly Bill 1177 (2014), available at http://www.leginfo.ca.gov/pub/13-14/bill/sen/sb_1151-1200/sb_1177_bill_20140828_enrolled.htm.

vi Policymakers might choose a timeframe shorter or longer than three days. FERPA currently requires the production of education records within 45 days, (34 C.F.R. § 99.10(b)), which is not sufficient for informing parents of their child's current educational progress, for providing them with timely information upon which to base decisions, or for helping parents identify any additional educational services their child might need.

INDEX

A

B

C

D

E

F

G

H

I

J

K

L

M

N

O

P

R

S

T

U

V

W

Y

Z

new literacies

AND DIGITAL EPISTEMOLOGIES

Colin Lankshear & Michele Knobel

General Editors

New literacies emerge and evolve apace as people from all walks of life engage with new technologies, shifting values and institutional change, and increasingly assume 'postmodern' orientations toward their everyday worlds. Despite many efforts to take account of such changes, educational institutions largely remain out of touch with the range of new ways of making and sharing meanings that increasingly mediate and shape the lives of the young people they teach and the futures they face. This series aims to explore some key dimensions of the changes occurring within social practices of literacy and the educational challenges they present, with a view to informing educational practice in helpful ways. It asks what are new literacies, how do they impact on life in schools, homes, communities, workplaces, sites of leisure, and other key settings of human cultural engagement, and what significance do new literacies have for how people learn and how they understand and construct knowledge. It aims to challenge established and 'official' ways of framing literacy, and to ask what it means for literacies to be powerful, effective, and enabling under current and foreseeable conditions. Collectively, the works in this series will help to reorient literacy debates and literacy education agendas.

For further information about the series and submitting manuscripts, please contact:

Michele Knobel & Colin Lankshear
Montclair State University
Dept. of Education and Human Services
3173 University Hall
Montclair, NJ 07043
michele@coatepec.net

To order other books in this series, please contact our Customer Service Department at:
(800) 770-LANG (within the U.S.)
(212) 647-7706 (outside the U.S.)
(212) 647-7707 FAX

Or browse online by series at:
www.peterlang.com

www.ingramcontent.com/pod-product-compliance
Ingram Content Group UK Ltd.
Pitfield, Milton Keynes, MK11 3LW, UK
UKHW021922120926
8556IPUK00042BA/1170

9 781433 128943